W9-AHR-882

WHO GOES THERE?

The Search for Intelligent Life in the Universe

EDWARD EDELSON

DOUBLEDAY & COMPANY, INC.
Garden City, New York
1979

Library of Congress Cataloging in Publication Data

Edelson, Edward, 1932–
Who goes there?

Includes index.
1. Life on other planets. I. Title.
QB54.E43 574.999
ISBN: 0-385-11257-2
Library of Congress Catalog Card Number 77-80882

Printed in the United States of America

First Edition

CONTENTS

LIST OF ILLUSTRATIONS

following page 52

following page 147

INTRODUCTION

There are some scientists who have a breathtakingly optimistic vision of the human race in the twenty-first century.

On the earth, science and technology have met the major challenges facing humanity. Famine has been held off; overpopulation has been checked; energy supplies are ample. Terrestrial affairs are run with the help of computers whose power is greater than that of the human brain. Humanity is extending its powers by symbiosis with machines and computers. Because of science and technology, the human race has achieved a level of well-being never before possible.

With its home base secure, humanity is advancing into space. There is a small but growing colony on the moon, where the techniques of lunar mining have been mastered. Several thousand people live in a self-contained space colony, and lunar ores are being processed to build more space islands. Mars has been thoroughly explored by unmanned rovers, and the first manned Martian colony is being planned. Scientific teams are working on long-term programs to transform Mars from a bleak and inhospitable desert to a planet that is suitable for human habitation.

Almost—but not quite—forgotten in this period of busy extraterrestrial growth is a decades-old project in which teams of scientists from every nation on earth are methodically searching for signals from other civilizations in the universe. It is a program that is being carried on both on earth and in outer space, using the earth's best minds and its most complex technology. Advanced computers working at the biggest astronomical facilities ever built are automatically scanning millions of radiofrequency channels from star after star. Hundreds of thousands of stars have been scanned over the decades. Aside from a faint twitch now and then, the results have been entirely negative.

Then, one day, the computers detect signs of organization in one of the channels they are scanning. The vision is that of Frank Drake, an astronomer at Cornell University who is one of the leading figures in the search for extraterrestrial life:

> . . . a signal is received from a direction fixed among the stars. Its frequency never changes. Thus, it must be of artificial origin. Its creator has corrected for the Doppler shift at the orbital motion of the object from which the signal comes. Only now is it noticed that every ten minutes the phase of the electrical vector which describes the signal is precisely inverted.
>
> This is absolutely conclusive evidence of the intelligent origin of the signal. The signal is carrying a message in binary code, a message that will be received for a year before the scientists understand the format of the message and recognize it as the song of people who have been alive, every one of them, for a billion years. They are sending the information which will make this same immortality possible for all the creatures of earth.
>
> As never before, the people of earth now hold their destiny in their own hands. . . .

That is the dream of the scientists who have created the field that is sometimes called exobiology and sometimes called SETI—the search for extraterrestrial intelligence. It is a vision of the human race enriched and even transformed by the wise use of technology and by contact with older and even wiser civilizations.

SETI is a young concept, but it is old enough to be suffering through its first crisis. The crisis is due to the close relationship between SETI and big, complex technology.

The idea of communication with civilizations elsewhere in the universe came of age in the 1950s and 1960s, a time when big technology was in its fullest flower. There seemed to be no limit to the growth of technology, especially space technology. Project Apollo had placed men on the moon. An ambitious program for reconnaissance of the solar system had been conceived and was to be largely carried out in not much more than a decade. Even more ambitious space programs were proposed.

Scientists in the field got used to thinking big, in billions of dollars and hundreds of millions of miles—not out of bravado, but because the vastness of the solar system and of the universe demands it. And, inevitably, the scientists who are working to make contact with other civilizations think big because their subject demands big technology.

The elementary calculations in SETI indicate that to make contact with another advanced civilization, we must either be very lucky or else use big machines for a long period of time. Modern technology may seem impressive. But if we compare the search for extraterrestrial civilizations to a poker game, the human race now has roughly the equivalent of one blue chip—a technology that barely gives us the ability to get into the game. Maybe we will be lucky enough to win the pot with our first ante. But the odds appear to be against us. The assumption in SETI is that we must start with today's technology and move up tenfold, a hundredfold, a thousandfold.

That sort of technological extrapolation was unchallenged in the 1960s, when all the arrows pointed up. But the 1970s have seen the rise of a different spirit. The 1960s were the decade of growth and expansion, of "bigger is better" and "the sky's the limit." The ideas that seem to have captured the public imagination in the 1970s are "small is beautiful" and "the limits to growth." A decade ago, only the benefits of big technology got consideration. Today, for an influential minority, at least, only the dangers of big technology seem evident, and there is a feeling that rather than exploring the universe, humanity would be better off tending its walled-in garden.

The scientists in SETI are keeping the technological faith, but they are very much aware that the issue is in doubt. In the complex formula by which the scientists attempt to estimate the number of advanced civilizations in our galaxy, perhaps the dominant factor is the lifetime of a technological society. One possibility is that a society such as ours will neither wipe itself out with nuclear weapons nor poison itself with pollutants, but will simply turn its back on technology.

Some scientists and philosophers of science say that European civilization, which first came to dominate the earth and then pushed into space, is unusual in the relatively brief history of human civilizations. We assume that the urge to explore and conquer is a basic attribute of human beings, but this may not be so. The Western idea of progress, which implies continued growth and expansion, and a continually increasing control over nature, is not the standard pattern on earth. Most civilizations have thought in terms of living in balance with nature, not of conquering it. Instead of a future of open-ended expansion, they have thought of recurring cycles, motion without progress. Instead of putting their hopes in the future and in progress, other civilizations have tended to look backward to a golden era of the past.

In the West, the idea of stasis is gaining ground. But the issue is by no means decided. Ten years from now, we may see the first years of space exploration as the promising but crude forerunner of true conquest of the universe. Or we may look back on our space adventure as an entertaining but essentially extraneous episode in humanity's earthbound future. The matter is being decided right now; not in any sweeping, grand debate, but in the more prosaic arena of budget hearings. Engineers, scientists, and other believers in SETI talk mostly to each other. So do the antitechnologists. Meanwhile, Congress (and its Soviet counterpart) wrestles with this great subject in terms of money. Do we spend the half-billion dollars or more needed to send rovers to Mars in the 1980s? Do we allow an appropriation for studying the feasibility of a multi-billion-dollar radio-telescope facility devoted to SETI? Will we spend the money for rockets that could let us build space colonies? Do we appropriate a few million dollars so that radio telescopes can begin a search for extraterrestrial signals?

It may seem absurd to reduce something as meaningful as SETI to haggling over money, but the method makes sense. By and large, Congress will do what Americans want it to do. If our faith in space exploration and its technology is alive, we will continue to push out into the solar system and beyond. If we turn thumbs down on technology, space exploration and the search for extraterrestrial intelligence will either limp along marginally or will fade away. If we give up, we save fairly large sums of money but lose the chance of being inducted into a galactic civilization whose capabilities could be beyond human understanding. Admittedly, it is a relatively slim chance, but those in SETI believe that it is worth the effort, even at great cost. Opposing that view is the thought that humanity should first see to its own basic needs, using the simplest possible technology, rather than building expensive space cathedrals while people starve.

The debate could be settled in a simple, catastrophic way. The big-technology option could be foreclosed because the earth suffers a global breakdown in any one of the many ways that pessimists can describe in vivid detail. We could starve ourselves out of existence, or upset the planetary balance and make the earth unlivable, or use all the nuclear weapons that have been perfected and distributed by political leaders. The survivors would then know with some certainty the expected lifetime of a technological civilization on earth. The knowledge would not do them much good, but it would settle a major point of debate in SETI.

Sometime in the next few decades, we will find out whether there is intelligent life on earth. At that point, the search for life elsewhere in the universe may begin in earnest. And here is the paradox: the fact that the search has begun here will be a good sign for its success. For we will then know that one technical society—ours—has begun an effort to contact others. That knowledge will give grounds for believing that the same decision has been made elsewhere in the universe.

For those who believe that humanity's future is tied to large-scale technology, the very existence of a search for extraterrestrial life will be a hopeful indicator for the human race. For them, the choice is between the earth and stagnation, or outer space and continued expansion. The question of spending a few million dollars for an organized, continuing search for extraterrestrial signals thus becomes one of the great decisions of our time.

If we pare away the externals, the issue seems clear-cut. But the externals, never are pared away. The decision about a search for signals from an extraterrestrial civilization will be made against a kaleidoscopic backdrop of conflicting interests: public interest in the technology of space, public demands for less federal spending, a fascination with unidentified flying objects by most nonscientists, skepticism about UFOs by most scientists, trust of scientists as the bringers of progress, distrust of scientists as the creators of technological monsters. And more.

It is hard to predict what the search for extraterrestrial intelligence will look like next year or ten years from now. But it can be said that the swift rise of this new discipline to scientific respectability is one of the more interesting phenomena of our time. Because of SETI, the human race can look at itself, its star, its planet, its solar system and its galaxy in a different light. Science has changed mankind's place in the universe drastically; from a species that bestrode creation, we have become just one product of evolutionary forces on a minor planet of a mediocre star. Now science offers us the hope of moving back toward galactic importance; if not on our own, at least in cooperation with other species who share an intelligent mastery of technology. To the pioneers in the search for extraterrestrial intelligence, this prospect is the most exciting one that the human race can have. Now we must see whether those pioneers can make their vision come true.

Already we have passed one milestone: we have gone off the earth to look for life on another planet. The Viking mission of 1976 sniffed around for life on Mars. Its findings, like most other data in SETI, were indefinite, provocative, and subject to debate. But the truly important point about the Viking mission was that it was accepted as a normal step in the exploration of space, even though the principles behind the mission would have been regarded as impractical dreams a few decades ago. Viking meant that exobiology is in the mainstream of science. The human race went a long way both physically and psychologically when life-detection instruments began operating on Mars. Viking is a good place to begin our search for life elsewhere in the universe, as well.

WHO GOES THERE?

CHAPTER ONE

VIKING TO MARS

THE LANDING

"Touchdown! We have touchdown!"

The time was 4:53 A.M. on July 20, 1976, the place, the Jet Propulsion Laboratory in Pasadena, California. The voice on the loudspeaker belonged to Richard A. Bender, the head of a group of engineers and scientists whose job was to see that a Viking spacecraft landed safely on the surface of Mars. He was a happy man.

Everyone at JPL that night was smiling. Viking I had landed perfectly, and people were cheering and jumping. The exultation was inevitable, since about seven years' effort and a billion dollars were riding on the landing. But there was an extra measure of exultation that came from the unusual suspense that was built into the final minutes of the descent from Martian orbit to the surface of the planet.

For almost twenty minutes, the scientists and engineers around the consoles at JPL knew as little about the fate of Viking as anyone else on earth. When the descent began, Mars was more than 220 million miles from earth. Radio signals from the spacecraft to earth could travel no faster than the speed of light. There was an unavoidable twenty-minute delay as the signals made the 220-

million-mile trip. If anything went wrong, JPL would not know about it for those twenty minutes.

They were the chanciest minutes of the mission. Viking was making what amounted to a blind landing on an unspecifiable spot in uncertain terrain. Indeed, the very fact that the landing was taking place on this date, July 20, was a tribute to the uncertainties of Mars.

The original date was July 4, 1976, a nice, round bicentennial day selected on what could be called the why-not principle: as long as Viking I (it was the first of a pair; the second was not far behind) had to land early in July, why not Independence Day in particular? But that date created a small deadline problem in verifying the safety of the landing site.

The primary landing site was on a Martian plain called Chryse, near the equator of Mars. It was chosen from careful study of the pictures taken by Mariner 9, which had orbited Mars for nearly a year, starting in November 1971. The Chryse landing site was picked because it seemed to combine interesting terrain with a good degree of safety. The site was at the mouth of a feature called Valles Marineris, a 3,000-mile-long canyon that could swallow up the Grand Canyon with ease. From the Mariner picture, the Chryse area appeared to be a drainage basin for a number of sinuous channels, which looked like dry riverbeds. If there had ever been running water on Mars—a question that was being discussed with intensity when the two Vikings were launched in August and September 1975, and all through their 400-million-mile voyages—Chryse would be a logical place to search for its effects. Since most scientists believe that water is essential to life, Viking would look for life in a place where water might once have flowed on Mars.

But the Mariner 9 pictures were on too large a scale to pinpoint a safe landing site. Their resolution was good enough to show large features, but they could not show smaller irregularities that would be enough to ruin a Viking attempt at landing. The final landing site could be selected only after the Viking spacecraft went into orbit around Mars. On board the Viking orbiter were cameras whose resolution was much greater than those carried by Mariner 9. As Viking circled Mars, its orbit would be adjusted to bring it over the proposed landing site.

The pictures transmitted from the Viking orbiter would be only part of the material used to make a final decision on the landing site. Readings from a radar facility on earth would be studied with equal care. Just about the time that Viking went into Martian orbit, the Chryse site would come into range of the earth's largest radio telescope, at Arecibo, Puerto Rico. Working as a radar set, the 1,000-foot antenna at Arecibo would be able to obtain extremely good information about the smoothness of the Chryse plain.

When the Viking orbiter pictures began to come in, they showed a scene that presented both opportunities and problems. Mostly, they showed clear evidence that water had once flowed on Mars, and in large amounts. There had been some debate about the origin of the sinuous channels in the Mariner 9 pictures. (See Figure 1.) Most scientists believed that the channels were cut by running water, but some held out for lava flows or wind action. There could be no doubt about the features in the Viking pictures, which showed much finer details than the Mariner images had provided. Water had flowed on Mars, all right. There had been a northerly flow that had cut riverbeds in the Martian soil, leaving behind eddy channels and "islands." In some parts of the channels, water had flowed violently enough to strip away the surface layer, creating a patch of scabland.

The existence of the channels raised a number of questions. When did the water flow, recently or long ago? Where had the water of Mars gone? Was the era of water flow on Mars a once-only event, or did the Martian climate go through cycles that would periodically bring floods to the now-dry rivers?

Those questions could be tackled later (although speculation began immediately). For the moment, the signs of water were good news. Water means life, or at least the possibility of life. One major objective of the Viking mission was to look for signs of life on Mars. Therefore, the indications that water had once flowed in the landing area meant that the search for life was off to a good start.

But the pictures also showed that the designated landing zone was not as safe as had been thought. The ancient Martian rivers had not emptied into the Chryse region; they had flowed right across it, cutting deep into its surface. The Viking scientists were looking at a region that was much more complex and much rougher than had been anticipated. The area could be studded with blocks big enough to tip the Viking lander over. If one of the lander's three legs should hit a large block, the craft could end up helpless on its back, like a turtle on earth.

Reluctantly, Viking officials gave up the July 4 landing date so that the area north of the original landing site could be reconnoitered. The Viking orbiter's cameras were pointed at the more northerly area. They showed smoother terrain and diminishing channels. The Arecibo radar signals crept across the area. They found a spot that appeared no rougher than average Martian terrain. The new site was chosen: 22.4° North and 47.5° West, by the latitude and longitude established by terrestrial scientists. (More accurately, the landing site was an elongated elliptical area centered on that point. There was a potential drift of several miles in the exact landing point.) And the new landing time was set.

Viking landed quite the way it was supposed to. The scientists could use the word "perfect." The lander separated from the orbiter; rockets were fired to start its descent; a protective cover was jettisoned. At the appropriate times, the lander's three legs were extended, a parachute was deployed to slow its fall, and a radar altimeter ignited three descent engines that set the craft down on the Martian surface with nothing more than a slight bump. In retrospect, it was all routine for a landing on Mars.

But think about seven years' work and a billion dollars spent on planning, building, testing and monitoring, all coming down to one moment. The pressroom and the auditorium at JPL were jammed with people who had stayed up all night. On the television monitors, you could see the men in the control rooms doing the silly things that come in the moment of jubilation. James S. Martin, Jr., the rock-solid project director of Viking, was wearing a funny T-shirt. Everyone was laughing.

A few minutes later, it was back to business. The computer on board the Viking I lander was programmed to begin taking pictures immediately after landing. One of the lander's two cameras was pointed toward a footpad of the spacecraft and the Martian soil around it. It was an unusual camera, one that did not have a lens. There was a moving part: a mirror that nodded up and down. Light from the mirror was captured by a sensor. The sensor translated the light signals into electronic signals, which were beamed back to earth.

At JPL, the electronic signals were translated back into light signals. As the mirror nodded, the entire camera rotated slowly, building up an image as a collection of vertical strips. (See Figure 2.)

THE PICTURES

You could call the camera a life-detection system. It could detect a Martian plant or an animal. The scientists had said jokingly that if anything walked by the spacecraft on Mars, its image would be sent back to earth—although it had been calculated that an elephant might escape detection if it walked just ahead of the swiveling camera. No one expected elephants, but it was hard to tell what to expect in the first picture from Mars. Indeed, although the camera system had been tested and retested exhaustively on earth, no one could be sure that it would work well on Mars, and there was uncertainty about the quality of the images that would arrive at Pasadena across 200 million miles of space.

The audience in the pressroom and the auditorium would find out as quickly as the scientists, since the picture was put on the television monitors as it came in. Thomas A. Mutch of Brown University, the leader of the lander imaging team, gave a running commentary. From the first, Mutch's straight scientific narrative was studded with superlatives: "incredible . . . fantastic . . . unbelievable . . ." He was talking about the quality of the pictures, which might have been taken just around the corner rather than 220 million miles away. The first picture from Mars, building slowly on the TV monitor, showed rivets in the cup-shaped footpad, Martian sand that had collected in the footpad, pebbles a fraction of an inch across, rocks half-buried in Martian sand. There had been those worries about the quality of the pictures that the lander cameras would take. As it turned out, the doubts were unnecessary. The camera system was superb.

The excellence of the system was proved in the second Viking picture, a 300° panorama of the Martian surface near the lander. This time, the content as well as the quality of the picture was reason for superlatives. In the time of testing, the camera system had been taken out into the desert of the American Southwest. The first Martian panorama showed a scene that could have been part of that desert. (See Figure 3.) The scene was impressive not because it was Martian but because it was so deceptively earthlike. If you looked hard, you could pick out an abandoned Volkswagen, a crushed beer can, a dry stream bed, and even a crooked letter "B" on the side of a rock. None of them were there, of course, but the impression of a terrestrial scene was hard to shake. Even the brightness of the sky helped create the impression. Before Viking landed, most Viking scientists assumed that the lighting on Mars would be more moonlike than earthlike, a glaring contrast of brightly lighted areas when the sun hit and deep shadows where it didn't. The earth's atmosphere carries suspended dust particles that scatter sunlight. The moon has no atmosphere, so light is not scattered. The Martian atmosphere is only a small fraction of the earth's. If we set up a unit of atmospheric density and say the earth's atmosphere rates at 1,000, the atmosphere of Mars rates at about 7—less than one percent as dense. (There is such a unit, called the millibar.) Therefore, Mars was not expected to have a bright sky.

But it does. Apparently, there are enough suspended dust particles in the thin Martian atmosphere to soften the glare. The first color picture, taken the next day, added an exotic touch: not only is Mars red, but its sky is pink. We can say quite accurately that Mars is rusty: its red color comes from iron oxide, or rust. The sky is pink because the dust in the Martian atmosphere is red, from those rusty rocks.

(One interesting sidelight: the color picture was made by superimposing three color images—red, blue, and green—then adjusting it until the overall color seems right. The first time around, the Martian sky was blue. After consulting their color reference standard, the imaging team made corrections and the sky became pink. The crowd of reporters booed loudly when the color change was announced—and got themselves criticized for "terrestrial chauvinism" by astronomer Carl Sagan.)

The first picture of Mars was a sensation. The first color picture was a marvel. The pictures kept coming, one after another. The scientists grabbed them for study, and the reporters' attention turned elsewhere: to the search for life, which everyone knew was the main purpose of the Viking mission.

The main life-detection experiments were to start on the eighth day after the landing. A surface sampler—a scoop on the end of an extendable arm—would reach out and pick up some soil. A sievelike lid would close over the sample. The arm would be pulled back over an inlet in the lander body. It would turn over, and the scoop would be shaken. Fine particles of Martian soil would fall into the inlet for analysis.

THE TESTS

What sort of analysis? Put it this way: Suppose you have the opportunity to land 160 pounds of instruments on a strange planet. You want to find out everything you can about that planet: all about its geology, its weather, its soil, and every other characteristic you can think of. There's a slim chance that something is alive on the planet. You don't know how big it is, or how much it differs from living things on earth. You don't have unlimited time, but you do have close-to-unlimited money to design instruments to detect that alive something. You know how to look for life on earth because the chemistry of terrestrial organisms is quite familiar. You don't know anything about the chemistry of life on the other planet, where nothing is quite like it is on earth. By what principle do you design your instruments to maximize the chance of finding that living thing?

The Viking scientists made what has been called the "assumption of mediocrity." It is an idea that runs like a red thread through all of SETI. In this instance, they assumed that life on earth is nothing special, but is so ordinary that life on another planet is likely to be pretty much the same. The starting point for biology on Mars was biology on earth.

On purely logical grounds, the assumption of mediocrity makes sense. If you don't know what you're looking for, look for something familiar. Terrestrial scientists have just one example of planetary biology, life on earth, so they start with that one. But there are also some good biochemical reasons for the assumption. There are only a limited number of ways in which atoms can be put together into molecules—at least those molecules associated with life on earth. Essentially, life requires long, complex chains of atoms. No other atom forms chains the way that carbon does. Organic chemistry is defined as the chemistry of carbon, and "organic" carries with it the connotation of life (although organic molecules can be produced without living organisms).

So one thing you do is to look for organic molecules. The Viking instrument for detecting organic molecules in the soil was the gas chromatograph mass spectrometer, which had other missions as well (analyzing the composition of the atmosphere, for instance). In the instrument, a sample of soil was first heated to nearly 400° Fahrenheit, boiling off gases and organic compounds. The compounds were then sorted out by mass and composition. The sample in the oven was then heated to 500° F., and the compounds that boiled off were sorted out again. Eventually, the instrument gave a good reading of the kinds of molecules to be found in the region around it.

At the beginning, it was assumed that the gas chromatograph mass spectrometer would be of only passing interest in the search for life on Mars. While the instrument could detect organic molecules, the presence of organic molecules in Martian soil would not indicate that the soil contained living things. Even a totally dead planet can collect organic molecules because they can be formed without the help of living organisms. Meteorites falling on earth contain fairly large concentrations of organic molecules. The same kind of meteorites fall on Mars, bringing organic molecules with them. So a positive result from the gas chromatograph mass spectrometer experiment would be far from conclusive.

Much more attention was focused on the life-detection package: three instruments packed into one cubic foot of space. Each instrument was designed to test a slightly different assumption about Martian life. Each of the tests was based on the principle that Martian life is chemically similar to terrestrial life, centering on carbon compounds. There was another basic principle: life on Mars, like life on earth, probably is based on water. It seems probable that some kind of liquid is needed by living organisms because chemical reactions go so well in solution. In theory, several liquids could support life. In practice, it seems difficult to picture a kind of life based on anything but water. Perhaps there are ammonia oceans on a distant planet, with strange fishes swimming in them, but our search now assumes that other planetary oceans consist of water.

One instrument on Viking, the pyrolytic-release experiment, assumed that a living Martian organism would absorb carbon-containing gases from the atmosphere. (See Figure 4.) After the sample of Martian soil dropped into the pyrolytic-release test chamber, it would be incubated for five days in a natural Martian atmosphere, to which would be added carbon dioxide and carbon monoxide, both labeled with radioactive carbon-14. A xenon arc lamp would provide artificial sunlight. After five days, all the gases would be flushed from the chamber, and the soil sample would be heated to vaporize whatever organic material might be present. The resulting vapors would be routed past a carbon-14 detector. Since all the other carbon-14 would have been flushed from the chamber (the reasoning went), any carbon-14 that remained would have to come from Martian organisms that had breathed the atmosphere and absorbed the test gases.

The pyrolytic-release experiment used no water. A second test, the labeled-release experiment, assumed that Martian organisms need some water to function. In this experiment, the sample of Martian soil was moistened with a small quantity of water containing organic molecules, again labeled with radioactive carbon 14. In essence, the experiment threw a piece of meat to the Martian organisms and waited to see if they burped. While the pyrolytic-release experiment would yield results only after the five-day incubation period, the labeled-release experiment would sniff constantly for gases given off by Martian organisms; there would be a reading once a day. The Martian soil sample would be incubated at 47° F. and would be followed for more than a week. A set of detectors in the chamber would make a continuing search for radioactive gases. If a microorganism fed on the nutrients and excreted carbon-containing gases such as carbon dioxide, the radioactivity of the atmosphere in the chamber would increase. The shape of the curve showing the increase would be quite important. If living organisms were excreting the gas, the curve would show continuing growth due to reproduction. A chemical reaction that produced the gas without the presence of living organisms would not show the growth typical of reproduction.

The third life-detection test, the gas-exchange experiment, did not use radioactively labeled compounds. As in the labeled-release experiment, the Martian soil sample would be wet. This time, the water would contain a rich broth of nutrients, nicknamed "chicken soup" by the Viking scientists. A sensitive gas chromatograph would monitor the chamber for traces of gases that might be emitted by a living organism which fed on the broth. At the beginning of the experiment, the water would be kept below the level of the soil sample. Later, the water level would rise to wet the soil. All the while, the gas chromatograph would search for oxygen, methane, carbon dioxide, and other gases that are associated with life on earth.

It was assumed that life on Mars, if it existed, would be tough, tenuous, and rare. The Viking scientists were not expecting herds of Martian animals to wander by, or fields of Martian plants. Sure there were signs of water flow on Mars, but that was a balmier time. Things are tough on Mars now, at least by terrestrial standards. Not only is the atmosphere thin, but it also is 95 percent carbon dioxide, with only a few traces of the water, oxygen, and nitrogen that are essential to life on earth. Mars is cold, too. The day that Viking landed, it recorded a temperature of −25° F. at Chryse, in a temperate part of the planet at a warm time of the year. In addition, the thin atmosphere of Mars does not screen out lethal ultraviolet radiation, which is absorbed by the ozone layer in the earth's atmosphere. Ultraviolet radiation is harmful to living organisms.

Even given all these handicaps, we can work out a scenario that would allow life to exist on Mars. Life could have evolved early, during the more favorable era when water flowed freely. As the water disappeared, living organisms could have slowly developed protective mechanisms; existing under the surface, hoarding the water that was available, and getting protection from ultraviolet radiation from the covering layer of soil. There are organisms on earth living under conditions of comparable severity—in boiling springs and the dry cold of the poles. Some of the Viking scientists were optimistic about life on Mars and some were pessimistic, but none of them really knew what to expect from the life-detection experiments.

After some initial trouble with the soil sampler (a pin that should have been released wasn't), the life-detection experiments began on schedule. The results were as dramatic as anyone could imagine. As soon as the first dose of nutrient solution was added to the soil sample in the labeled-release experiment, the detector picked up a steep increase in radioactivity. Gases that included radioactive carbon from the nutrients were being released at about four times the rate that would have been expected from soil on earth. The gush of gas did not last long; it stopped after the first 70 hours. When some more nutrient solution was added, there was another brief spurt of gas release, followed by another drop in activity.

As for the pyrolytic-release experiment, Norman H. Horowitz of the California Institute of Technology, who designed the instrument, expected a reading of about 15 counts per minute from barren, lifeless soil. The actual reading was 96 counts per minute. That was a higher reading than had been obtained in test runs with soil from the dry valleys of Antarctica, which are believed to be the most barren places on earth.

The results of the gas-exchange experiment were also unusual. In the first stage of the experiment, when the nutrient solution was not in contact with the soil sample, there was a quick burst of oxygen—fifteen times the amount that could be expected from soil on earth—and a slow, steady release of carbon dioxide. When the level of the solution was raised to wet the soil, no more oxygen was released, and the amount of carbon dioxide actually dropped.

The Viking scientists circled cautiously around those results, trying to make some sense out of them. Basically, there are two competing explanations for the Viking test results. They could be caused by some sort of chemical reaction that has nothing to do with living organisms. Or they could be caused by living organisms in the Martian soil. Chemistry or biology: the scientists had to choose one or the other.

Between the day that Viking I landed and the day in May 1977, when the life-detection experiments on both Viking landers were turned off, those experiments were run again and again in an effort to get a clear-cut answer. Sometimes the soil was heated to several hundred degrees to see whether this kind of heat sterilization would affect the response. Sometimes heat sterilization at lower temperatures was tried. On some runs of the pyrolytic-release experiment (the dry one), water was added to see how its presence would affect the results. The Viking II lander, which put down safely at a more northerly and presumably wetter region of Mars, went through the same round of experiments. Working from earth, the Viking scientists dug a foot-deep trench to get a soil sample from deeper down, where ultraviolet radiation could not penetrate. (See Figure 5.) They shoved a rock aside to get a soil sample. They ran shorter and longer cycles of the experiments under a number of different conditions, to try to end the uncertainty. In the end, a degree of uncertainty remained.

Consider one experiment, done with the pyrolytic-release apparatus on Viking I. Team leader Norman Horowitz first moistened the soil sample and then heated the sample to drive off the water. By then, early in 1977, it had been established that either adding water or heating reduced the response. Doing both, Horowitz reasoned, should lower the response to zero.

Instead, there was a significant radioactivity count from the sample. That result seemed to be a vote for a chemical cause, rather than biology. Water is assumed to be essential to life. Yet adding water to the Martian soil sample reduced the response, while removal of the water by evaporation increased it. In addition, heating would presumably damage Martian microbes; yet there was a high level of activity after the heating.

But there were the results of the gas-exchange experiment, in which Martian soil that was exposed to water vapor gave off a lot of oxygen quickly. Adding water could promote the release of oxygen without the presence of living organisms if the Martian soil is rich in oxygen-containing compounds. The fact that there is little or no oxygen release if more water is added a few days later supports the chemical explanation: after the initial chemical reaction, there was no more oxygen to release.

It seems simple and consistent. But some scientists, such as Bernard Levin, head of the gas-exchange experiment team, detected inconsistencies. If all the responses were due to reactions of water with oxygen-containing compounds, Levin said, there should have been no activity in the pyrolytic-release experiment when the soil was first wetted and then heated. If the activity was due to chemistry, Levin said, then Martian chemistry is considerably more complicated than terrestrial soil chemistry.

After a while, the weight of opinion began to tip toward the chemical explanation. Taken alone, the three life-detection experiments might be considered to have given equivocal results. But the readings from the gas chromatograph mass spectrometer carried a great deal of weight. That instrument looked repeatedly for organic molecules on Mars. It found none—none at all, an astounding result.

WHERE HAVE ALL THE ORGANICS GONE?

In effect, the gas chromatograph mass spectrometer was looking for bodies. On earth, soil is rich with the remains of microbes which have lived and died, leaving their organic material to mark their existence. There are no such bodies in Martian soil—at least as far as the gas chromatograph mass spectrometer can tell. And it is a sensitive instrument.

"Our instrument could have detected dead organisms anywhere from a hundred to one thousand times fewer in number than there are in poor earth soils, like the soils of Antarctica," said Klaus Bierman of the Massachusetts Institute of Technology, the designer of the instrument. "We did not find any organics. There doesn't seem to be a mechanism that accumulates organics on Mars."

The puzzling thing about the Martian organics is their complete absence. As stated earlier, organic molecules do not have to come from living organisms. Carbon-rich meteorites contain organic molecules that have formed naturally, and these meteorites have rained down on Mars for eons. Where have these organics gone?

A number of theories have been floated to explain the missing organics. They could have been destroyed in some way by the huge dust storms that sweep across the face of Mars. Or they could have been eaten by Martian microbes. After all, Mars is a poor planet by terrestrial standards. If it does have microbes, they might not have the luxury of being left alone after death. Martian microorganisms may have developed into efficient cannibals under the pressure of the harsh Martian environment. Or they may have developed hard shells to lock in water and keep out radiation, shells that would make them undetectable.

Many of the Viking scientists now believe that the missing organics were oxidized. The thin Martian atmosphere allows ultraviolet radiation to bathe the surface. Ultraviolet breaks Martian water into hydrogen and oxygen. The oxygen combines chemically with the organic molecules, destroying them. So the theory goes.

NO LIFE ON MARS?

When Viking project scientists met in Boston in September 1977, the consensus was that life had not been found on Mars. If there had been a vote at the meeting, the chemical explanation would have won. The chemistry of Mars may be complicated, but the assumption is that the Viking results can best be explained by chemical reactions with a soil rich in oxygen-containing compounds.

Of course, we could say that if the chemistry of Mars is unusual, its biology is also likely to be unusual. But all things being equal, scientists accept the explanation that challenges the fewest accepted concepts. The concept of strange Martian soil chemistry is easier to defend than the concept of a strange Martian biology. In addition, conservatism is important in exobiology, which has climbed to scientific respectability only in the past few decades. The conservative interpretation of the Viking results is the chemical interpretation. No one can deny that there is a good amount of data to support that interpretation.

Norman Horowitz pointed out that the readings from his pyrolytic-release experiments which seemed to indicate the presence of life had continued even after the soil samples were sterilized by heating. Horowitz said that the failure of heat sterilization to stop the release of radioactive gases was a telling sign against the biological explanation. If Martian organisms do exist, Horowitz said, they have been bred by evolutionary pressure to survive in the cold environment of Mars. Therefore, Martian microbes are presumably much more sensitive to heat than terrestrial microbes, and their activity should have been stopped by the intense heat-sterilization process. The fact that activity had continued after heat sterilization showed that the results of the pyrolytic-release experiment "couldn't be biological," Horowitz said.

Vance I. Oyama, the scientist in charge of the gas-exchange experiment, had the same phenomenon to report and the same interpretation. Activity had continued even after the Martian soil sample had been baked enough to kill terrestrial microbes. The fact that the observed reaction was "insensitive to heat sterilization tends to reduce the probability that it is biology," Oyama said.

But the vote for the chemical explanation was not unanimous. Aside from the fact that scientific issues are not settled by vote, belief in a possible biological explanation was kept alive by a few scientists in the Viking Project. One of them was Gilbert Levin. Levin said that he had tried to reproduce the Martian results on earth using many different oxygen-containing compounds of the kind that were supposed to be on Mars but "could not find equivalent reactions." Levin has been one of the Viking scientists who is most strongly inclined to the biological explanation. (Horowitz, by contrast, is one of the most determined skeptics. When, during the early days of the mission, he wavered slightly toward the thought that there might be life on Mars, the stock of the biological explanation went up a few points. Later, Horowitz went back to his original position.) Now Levin was saying that the labeled-release experiment had produced the same results on both landers at two different Martian locations. To him, those results seemed to be interpretable as "unequivocal evidence of life."

But the others kept coming back to the gas chromatograph mass spectrometer. Where are the organic compounds scientists would expect from living organisms? Their absence seemed to be the clincher for the chemical explanation.

But that doesn't close the case entirely. The maddening thing about the Viking results is that they have not really settled the question of life on Mars. All the scientists are talking in terms of probabilities, not certainties. Oyama and Horowitz believe that it is highly improbable that life exists on Mars. Levin believes that the possibility is still open. In Scottish jurisprudence, there is a verdict that falls somewhere between guilty and not guilty: not proven. That's not a bad verdict for the Viking mission, since the chemical explanation presents almost as many difficulties as the biological explanation. Maybe the best summary is a statement that Horowitz made in talking to an interviewer: "There's no question that there is something going on that we don't understand."

1984: BACK TO MARS?

The only way to get a good answer is to go back to Mars, which is something the scientists would love to do. Money is the problem. When, early in 1977, the National Aeronautics and Space Administration's budget was revised to include a bare $10 million to keep the idea of future missions to Mars alive, the Viking scientists reacted with a sort of shocked jubilation. They had expected to be shut out entirely. At the Jet Propulsion Laboratory, planning began for a Mars mission in 1984, just about the first realistic opportunity. The 1984 mission would have two landers, each with a vehicle that could carry laboratories for 100 miles or so. (See Figure 6.) In addition, the spacecraft would also carry penetrators, which would rather resemble giant thumbtacks. The penetrator would slam hard onto the surface of Mars. The point of the tack would shoot about ten yards into the ground. It would be connected to the flat head of the "tack" by a wire that would carry data about conditions underground. The data would be radioed back to the orbiting part of the spacecraft.

The rover-penetrator mission would be the prelude to a later voyage whose primary mission would be to bring back a sample of Martian soil for analysis—not back to the surface of the earth, but to an orbit around earth where scientists could carry out the analysis without fear of contaminating our planet with any microorganisms that might be in the Martian sample. The proposed 1984 mission would cost an estimated $700 million in 1977 dollars, and more when inflation is taken into account. The soil-sample-return mission would cost in the neighborhood of $1 billion.

Neither mission will fly, at least for the foreseeable future. The joy of the Viking scientists over the 1977 appropriation to plan a follow-up mission turned out to be premature. By 1978 it was clear that the Carter administration was not in favor of an expensive return mission to Mars. The $30 million to $40 million that would be needed to start planning for a 1984 Mars mission was not in the budget that the administration submitted to Congress in January 1978. More than that, it was clear that the Carter administration was not making any large-scale plans for planetary exploration. Money would be doled out, enough to keep the program alive, but spending would be grudging and every penny would be pinched. At least, that was the impression that space scientists received about the administration's attitude.

There was, of course, very little grumbling about the Mars mission decision, at least in public. On a practical level, grumbling would do no good. Neither the administration nor Congress was in a spending mood for space programs. Rather than pushing for bigger and better missions in space, scientists were trying to prevent existing programs from being cut back. The open-ended, liberal spirit of the 1960s had given way to the more cramped spirit of the 1970s.

But the scientific reaction was also attributable to the results of the Viking mission. Those results seemed to indicate that there is only a slim chance, at best, that life exists on Mars. If there had been a scientific consensus that life on Mars is probable, the pressure for a follow-up mission to Viking would be much greater. As it is, space scientists are not exerting that pressure. An ambitious new mission to Mars is a tempting prospect, but such a mission does not make practical sense if the chances of dramatic findings are small. Martian life would be the most dramatic finding of all—probably the only finding that could persuade Congress and the administration that a billion-dollar mission should be undertaken. The unwillingness of the scientific community to press for a Martian follow-up mission speaks quite clearly about the prevailing assessment of the Viking mission. By not asking urgently for money, the scientists are implying that Viking did not find life on Mars.

But this does not mean that planetary scientists have given up all plans for going back to Mars. Sooner or later, a new Mars mission will be proposed with enthusiasm. Planetary geologists still want to study Mars as a planet; meteorologists want to study its weather more closely. And Mars missions also can expand the slender data base on which exobiology is founded.

Exobiologists have only one planetary system to work with. They have assumptions about the probability of planetary systems around other stars and about life in such planetary systems, but they have only one system in which life is known to exist. In fact, they have only one *planet* on which life is known to exist.

By current standards, Mars is a borderline case for life. If there should happen to be living organisms on Mars, proof of their existence would change the complex chain of calculations that exobiologists use to estimate our chances of making contact with another intelligent civilization. The changes might not be profound, but they would be significant. If there is no life on Mars, the calculations would also be influenced. Even though the consensus is that Mars is lifeless, that belief falls short of certainty. And scientists crave certainty.

But suppose that against all the odds there is life on Mars. It could be that Martian organisms have an unexpected biological twist, some evolutionary oddity that indicates the existence of an unearthly pathway by which life can evolve. That knowledge would also change the odds for the existence of life on other planets around distant stars. It would, in fact, revolutionize exobiology, and it would add evidence to support a long-term search for signals from extraterrestrial civilizations. The belief that we will go back to Mars, sooner or later, is hard to resist.

CHAPTER TWO

WHERE WE COME FROM

Viking went to Mars because of a human vision of what Mars is like. You cannot plan to visit a place unless you believe that it is, indeed, a place. In the beginning, Mars and the other planets were things, not places. They were only points of light in the heavens. (The planets got their name, Greek for "wanderers," because they are moving points of light among the fixed points of the stars.) The earth was the only place. Later, in the youth and adolescence of modern science, the planets did become places. The pendulum swung far. The planets not only were places, they were also earthlike places, very easy to get to and certainly habitable—if not by humans then by beings similar to humans.

We can trace an intellectual path from the early views of the earth and the universe through what now seem to be the romantic days of science and on to today's rigorous world-view. But we should remember that this road is built by working backwards. The natural philosophers of the past could not appreciate that their ideas would someday be judged by people with our view of the universe. Modern scientists pick and choose among the ideas expressed over millennia, winnowing out all but those few concepts that fit the modern world-view. As Jorge Luis Borges said in his essay on Kafka's precursors, "The fact is that every writer *creates* his own precursors. His work modifies our conception of the past, as it will modify the future." In the same way, modern science continually modifies its judgments of the past as new concepts arise.

The major effort of modern science is to banish wonder from the universe. (Chance cannot be banished because quantum mechanics says that the universe is unpredictable on the most basic level.) In the beginning, the world was full of wonder. The earth was an almost unimaginably big place set in the middle of a universe that was almost beyond understanding. If anyone thought about it, the earth was regarded as the sole repository of life in a universe consisting of dots of light circling in complex patterns.

BEGINNINGS

But from today's vantage point, we can find the beginnings of a belief in extraterrestrial life at the very start of Western science, which seems to have begun in Greece. The first great school of ancient Greek science arose in Ionia, which occupied a narrow coastal strip of what is now western Turkey and the neighboring Aegean Islands. Miletus, one of the twelve major cities of Ionia, was the home of the first great natural philosophers. The first of the Milesian philosophers was Thales, who lived in the sixth century B.C. Anaximander, who followed Thales, entertained the idea that there might be other worlds than the earth. And in the fourth century B.C., the Epicurean philosopher Metrodorus wrote, "To consider Earth as the only populated world is as absurd as to assert that in an entire field sown with millet only one grain will grow."

That statement could stand as a succinct summary of today's most prevalent view about life in the universe. However, it was more a passing thought than a central tenet of Greek natural philosophy. A major interest of the ancient Greek intellectuals was the effort to organize the universe along severely logical lines. Pythagoras organized the world by numbers: square numbers, triangular numbers, pentagonal numbers, in ordered relationships. The stars and the planets circled the earth in perfect circles, another orderly relationship.

There was a fire at the center of the universe. Around that fire circled the earth, the moon, the sun, the five known planets, and the stars. When observations indicated that the orbits of the planets were not perfect circles, a scheme was developed in which the orbits were described as a system of smaller circles around larger circles. Cycle and epicycle; postulate enough of them, and any motion of any heavenly body fits the theory.

From our vantage point, we can pick out the dissenters from this logical but incorrect world-view. In the third century B.C., Heraclides of Pontus argued that the observations could be explained by saying that the earth is rotating on its axis. Aristarchus of Samos, who was born about 75 years after Heraclides, published a book, long lost, which included the hypothesis that the stars and the sun do not move, but that the earth orbits the sun.

At the time, the heliocentric theory was a minor curiosity. The world listened to Aristotle, who explained the case for a geocentric universe quite logically: since the gods are eternal, the heavens must be eternal; the heavens were a rotating sphere, with the earth at the center of the sphere.

Some 1,800 years later, Copernicus was aware that his heliocentric theory was a revival of Aristarchus's hypothesis. For most of that long period, the heliocentric theory was pushed out of sight. The stubborn faith in the belief that the earth is the center of the universe can be explained only partly by the technical difficulties of working out a heliocentric theory. (The planets move in ellipses, not circles, which complicates the calculations.) The real problem was an inability to grasp the nature and the size, of the universe. It is a problem that still bedevils SETI. Most nonscientific people, and some scientists, cannot comprehend the vast distances of the universe.

Looking out at the twinkling stars, it is hard to realize that if you started for the nearest of them using the most advanced machines now available, you would die long before you made even a decent start. Even at the ultimate speed, the speed of light (and of all other electromagnetic radiation, including radio waves), signals take decades or centuries to reach our nearest neighbors in space. Beyond those neighbors, the universe really begins. If man is the measure of all things, the universe is too vast to comprehend. We can understand it only by adopting a frame of reference that is quite literally inhuman.

A UNIVERSE TOO VAST TO COMPREHEND

Such a frame of reference is not easy for the human mind to adopt, even in an age which is impregnated with the measurements of modern technology. It was simply impossible to even the keenest intelligence of a nontechnological civilization. Most of Aristarchus's works are lost (his reference to the heliocentric system survives in a work by Archimedes), but we do have a book in which he attempted to work out the sizes of the sun and the moon. On the basis of his calculations, he said that the sun is eighteen to twenty times more distant from the earth than the moon, and that the diameter of the sun is between eighteen and twenty times greater than the diameter of the moon. The distances and sizes that Aristarchus described may have seemed enormous to him. In fact, his estimates fell short by many thousandfold. The flaw was not in the methods that Aristarchus used. (Not long afterward, another Greek scientist, Eratosthenes, used similar methods to make a remarkably accurate estimate of the size of the earth.) More likely, Aristarchus simply was not ready to deal with a solar system of huge bodies and vast distances. The universe was growing larger, but it was still a small, intimate place by modern standards.

Here and there in Hellenistic and Roman literature, the concept of life elsewhere in the universe is mentioned. In the first century B.C., Lucretius wrote that "it is in the highest degree unlikely that this earth is the only one to have been created" and that "in other regions there are other earths and various tribes of men and breeds of beasts." Three centuries later, Plutarch speculated about the suitability of the moon for life. Again, the thought is of human life in a universe built to human scale. The universe was a comfortably manageable home.

When Rome fell, the universe became even smaller. After the initial chaos of the barbarian conquests, a lively intellectual life was established in Europe. Medieval times were not the "dark ages" we were taught about in grammar school. Medieval art and the architecture that culminated in the great Gothic cathedrals tell their plain story of intellectual vigor. But it was not the questing, outward-looking vigor of the inquisitive scientific mind. A closed, stable society seemed best. A plurality of worlds would be an embarrassment, especially in theology. Earth had its Savior. Other inhabited planets would either require some attention from our Savior or a plurality of Saviors. It was an uncomfortable choice, best avoided by thinking only of Earth.

At this point in the conventional history, there is a description of Europe's awakening from its intellectual lethargy to resume its march toward modern science and technology. In the accepted wisdom, medieval times were an unfortunate lapse from the normal state of humanity, which features a continual striving for economic growth and control over the physical world. But as we have seen, it is the restless spirit of modern science that appears to be unusual in human history. For some reason, a few people began looking at the world in a different way about 400 years ago. The tools they had for looking at the world were in no way new or different. Those tools would now be used to weigh, measure, and conquer nature in a way that had not been thought of before.

GALILEO'S INTELLECTUAL GIANT STEP

The single figure who is most identified with the birth of modern science is Galileo Galilei. Some predecessors of Galileo had chipped away at the foundations of the medieval system. In the middle of the sixteenth century, Nicolaus Copernicus revived and expanded the concept of the heliocentric system, removing the earth from the center of the universe. Not long afterward, Johannes Kepler added the important information that the planets orbit the sun in elliptical, not circular, paths. But both Copernicus and Kepler were deeply rooted in the medieval way of thinking. Copernicus believed in the systems of cycles and epicycles, heavenly spheres within spheres. Kepler conceived of a logically organized solar system in which the orbit of each planet was related to one of the regular solids—the cube, the octahedron, and so on. Kepler also tried to relate the orbit of each planet to musical notes, so that there would literally be music of the spheres.

Galileo took the giant intellectual step of looking at the heavens without such preconceptions. Anyone could have done the same thing, just as anyone could have started modern technology. The Romans had the steam engine and the ancient Chinese had the rocket. They treated them as toys.

In the same way, peddlers were hawking telescopes in the streets when Galileo decided that these gadgets were something more than toys. The available instruments were not up to his standards, so Galileo made his own. He turned the telescope on the moon and saw mountains and craters. He turned it on the Milky Way and saw it resolve itself into individual stars. He turned it on Jupiter and detected four moons—the Galilean satellites, they are still called. He turned it on Venus and found that the planet showed phases like the moon. He turned it on the sun and saw sunspots, evidence that the sun rotates.

The evidence that the bodies in the sky are places, not abstract points of light, was there for anyone to see, and the evidence for the heliocentric theory was overwhelming to anyone who had eyes to look. Most scientists chose not to look. There is a celebrated story of Galileo offering his telescope to a critic, so that he could convince himself. The offer was declined because the medieval mind did not regard the physical world as the most important factor. There was a higher truth, dictated by logic and faith. Galileo's heresy, for which he was condemned to penance, was to challenge this higher truth.

A new spirit was abroad. Even before Galileo made his astronomical observations, Giordano Bruno, a Dominican monk, could write, "Innumerable suns exist. Innumerable earths revolve about these suns in a manner similar to the way the planets revolve around the sun. Living beings inhabit these worlds." Bruno was burned at the stake for heresy, not so much because he championed the Copernican theory as because of his fierce independence and refusal to compromise.

It was the beginning of a period of florid speculation about the universe (although, it should be noted, the word "universe" has a far different meaning to our infinity-minded age than it did when things were on a smaller scale). The old bonds imposed by religion were broken, and the new, equally strict bonds of modern science had not yet been forged. Imagination was free to run loose. Christiaan Huygens, a Dutch astronomer who discovered the rings of Saturn and a new satellite of Jupiter which he called Titan (we will hear more of it later), writing toward the end of the seventeenth century, concluded quite matter-of-factly that there are planets circling the stars, even though the stars were too distant for their planets to be seen from earth. It is equally probable, Huygens went on "that these great and noble bodies have somewhat or other growing and living upon them, though very different from what we see and enjoy here." Nobody burned Huygens or forced him to recant. Times were changing.

Cyrano de Bergerac wrote about a voyage to the moon, although he was unscientific about the mode of propulsion. Bernard de Fontenelle wrote in the eighteenth century about the inhabitants of other planets in the solar system. Two of the greatest English scientists, Isaac Newton and William Herschel, believed that the sun was inhabited.

THE MOON HOAX

By the middle of the nineteenth century, there was a more or less casual acceptance of the possibility that life existed on other planets in the solar system. That was proven by the Moon Hoax of 1835. The *New York Sun* printed a series of reports, allegedly written by Herschel, describing in great detail observations of living creatures on the moon—lunar mountains embedded with sapphires, rubies, and amethysts, moon creatures eating melons, and the like. The hoax was soon exploded, but the point is that people were ready to listen and believe.

Mostly, they were ready to listen to stories about Mars. Volumes could be—and have been—written about the fascination that Mars has held for believers in extraterrestrial life. William K. Hartmann and Odell Raper, in an official NASA history of the Mariner 9 mission, described "the conception of Mars that was transmitted by astronomers to other intellectuals of the early 1800's:"

> Mars was a planet with oceans and lakes, dry reddish land, clouds, polar snows, a day of some 24 hours—in short, it was a planet much like earth. In view of the momentous import attached by us moderns to the search for life in the universe, it

may come as some surprise to realize that Herschel and other scientists of his time almost casually assumed that Mars and other planetary bodies were inhabited. This was before Darwin gave us the idea of the long struggle toward sentient life; thus scientists of the nineteenth century inherited the idea of "the plurality of worlds," with a Mars already teeming with creatures and a reasonably pleasant environment well suited for them.

Much of the later story of Mars centers on one Italian word: *canali,* which can be translated either as "channels" or "canals." The word was first used in 1869 by Father Pietro Angelo Secchi, who published the earliest-known color sketches of Martian features. Father Secchi described some streaky Martian markings as *canali.* It is not known whether he wanted to say "channels" or whether he thought that there were canals on Mars. Nine years later, another Italian astronomer, Giovanni Schiaparelli, used the same word (and its singular, *canale*) to describe a network of lines which he said he had observed on Mars. Schiaparelli's map of Mars, showing the *canali* clearly outlined, set off a controversy that lasted the better part of a century.

Some astronomers said they could see the *canali;* some said they could not. There was the question of what could be seen when the viewing was excellent and what could be seen on those infrequent occasions when the viewing was better than excellent. Skepticism abounded. "I can't believe in the canals as Schiaparelli draws them," wrote E. E. Barnard, an American astronomer with an excellent reputation for observation, in 1894. ". . . I verily believe—for all the verifications—that the canals as depicted by Schiaparelli are a fallacy and that they will be so proved before many favorable oppositions are past." (An opposition is the time when the earth and Mars are closest together in space.)

THE ROMANCE OF MARS

Then along came Percival Lowell, of the Boston Lowells, who began observing Mars from an observatory he founded in Flagstaff, Arizona, toward the end of the nineteenth century. If Barnard could not see the canals, Lowell certainly could, better than anyone else has ever managed to see them. And he could not only observe the canals, built with "wonderful directness," but he could also explain the reason for their existence. Mars was an old, dying planet that was losing its water. Its once-prosperous inhabitants had become engineers on a huge scale:

"Irrigation, and upon as vast a scale as possible, must be the all-engrossing Martian pursuit. So much is directly deducible from what we learned at Flagstaff of the physical condition of the planet."

Lowell's vivid picture of a dying Martian civilization made a lasting impression on the public. It gave birth to one book of permanent value, H. G. Wells's *The War of the Worlds,* in which the earth was invaded by Martians who were bent on conquering our planet because theirs was becoming uninhabitable. In Wells's story, Martian technology was too advanced to be fought off by mankind, but the Martians were defeated when they died of terrestrial infections. Some forty years later, Orson Welles demonstrated just how strongly people believed that there might be intelligent life on Mars. His eerie radio dramatization of Wells's story set off a nationwide panic.

If the public believed in Martian civilizations, scientists did not. Evidence against Lowell's thesis was accumulating rapidly. At every Martian opposition, telescopes were turned on the red planet. Most of the observers saw nothing resembling an organized network of canals. Instead, they saw what one astronomer described as "a prodigious and bewildering amount of sharp or diffused natural, irregular detail . . . it was at once obvious that the geometrical network of single or double canals discovered by Schiaparelli was a gross illusion."

Other measurements of Mars slowly destroyed the belief that it was a twin of earth. In the 1920s, temperature measurements made from earth found that the mean Martian temperature is —40° F., which would make life on Mars either quite difficult or impossible for advanced organisms. In the 1920s and 1930s, another series of observations found no detectable water vapor or oxygen in the Martian atmosphere, another telling point against the existence of a Martian civilization. Gradually, a new picture of Mars emerged, a picture of a cold, almost waterless, almost oxygenless planet with only a wisp of atmosphere. The romance of Martian civilizations died as the scientific data accumulated. The possibility of life on Mars remained open, but the rules of the game had changed.

To scientists it was increasingly clear that if there was life on Mars, the organisms would have to be quite hardy by terrestrial standards and probably had to struggle even for minimal survival. It was a picture that the public, even some educated segments of it, failed to accept. During the opposition of 1924, the United States Army and Navy actually ordered their most powerful radio stations to stop sending for a time, to help in a scheme to detect possible messages from Mars. There were none, of course, but Martians made juicy reading, and their supposed civilizations were a staple of the tabloids and pulp magazines of the 1930s. In 1938, a Canadian astronomer, Peter M. Millikan, began a scientific paper with the weary observation: ". . . so much nonsense has been written about the planet in various branches of literary endeavor, that it is easy to forget that Mars is still the object of serious scientific investigation."

The evolution of the human view of Mars is instructive. First Mars was just a wandering light in the sky. Then it became a place, a globe like the earth—so much like the earth that we peopled it with beings like ourselves and a civilization like ours. Then, slowly, scientific studies turned Mars into what it is now: a planet that has some resemblances to earth, but is different in a number of basic ways, and one that certainly cannot support anything resembling human life-forms. Our current view of Mars is clear, accurate, and rather bleak.

While concepts of Mars were changing, so were concepts of the universe, and in much the same way. The universe was growing larger and more unhuman. In the beginning, it was just a collection of lights in the sky. When astronomers realized that the stars were not merely points of light but instead are bodies like the sun, they began trying to determine the distances to the stars.

MEASURING THE UNIVERSE

And just as mankind had imagined that Mars must be a homey, earthlike place, astronomers believed that the stars could not be all that distant. It seemed natural to believe that the universe was built to a human scale. It was only as measurements became more accurate that the vast, unhuman dimensions of the universe emerged. Even when the measurements were made, there was a major controversy about the size of the universe, a controversy due largely to the inability of even trained scientists to understand just how small the human race, its planet, and its sun are in the cosmic scheme of things.

One way to judge the distance to a star is by its brightness. If a star appears very bright, it seems reasonable to assume that it is closer to earth than a dim star. But this technique works only if all stars are identical. They are not. Some stars glow dimly. Others blaze brightly, burning with a blue-white heat. A bright star that is far away and a dim star that is close by will have the same apparent brightness.

A method other than judging the brightness of stars is to measure the movement of a star in relation to other stars. In reality, it is not the star that is moving, but the earth. In its orbit around the sun, the earth travels a fairly large distance. A nearby star seen from the earth appears to be moving against a background of more distant stars. The apparent shift of a star due to the earth's motion is called the star's parallax. In 1838, German astronomer Friedrich Wilhelm Bessel made the first measurement of parallax to determine the distance of a star from the earth. Other measurements followed, but the use of the method was limited because it works well only at fairly short range. The more distant a star, the less its apparent motion. Beyond a distance of thirty light-years, parallax measurements become more and more uncertain. Only a few hundred stars out of the myriad visible from earth can be placed by parallax measurements.

During the nineteenth century, astronomers used a combination of parallax measurements and brightness measurements to determine the distance of a number of stars. The method, although tedious, was successful because a classification of stars by type had been worked out. An astronomer would determine the distance from the earth of a star of a given type by measuring its parallax. The astronomer would then find a star of the same type that was far more distant and would compare its apparent brightness with the brightness of the star of known distance. The comparison of apparent brightness with actual brightness would give the distance of the star.

The task of determining distances would be infinitely easier if there were some characteristic of a star that was directly related to its brightness. Given such a relationship, the astronomer could simply measure the characteristic, determine the actual brightness of the star, and so place it accurately.

Such a characteristic was discovered early in the twentieth century by Henrietta S. Leavitt, an assistant at the Harvard Observatory. The observatory was making a systematic search for variable stars—so called because their light grows dim and then bright in a regular cycle. Miss Leavitt was studying a class called Cepheid variables, so named because one of the most noted is in the constellation Cepheus. She found that the period of fluctuation in a Cepheid variable's light output is directly related to its brightness: the brighter they are, the slower the cycle from dimness to brightness. By timing the cycle, an astronomer could determine the actual brightness of a Cepheid variable. By comparing the apparent brightness to the actual brightness, the distance of the star could be determined.

Miss Leavitt's discovery came along at a critical time in a long-standing debate about the structure of the universe. The debate goes back at least as far as the Greeks. On cloudless nights, they could easily see a band of light running across the sky. They called it *galaxias kyklos,* the "milky circle." We call it the Milky Way.

In the fifth century B.C., Democritus wrote that the Milky Way might be made of stars. Galileo found that it was when he trained his telescope on it. A little more than a century later, an English philosopher named Thomas Wright wrote a book theorizing that the stars in the Milky Way were in a galaxy shaped like a grindstone, with our sun near the center. Wright added that our galaxy probably was only one of many—an idea that was echoed by Immanuel Kant just a few years later.

Wright's book caught the attention of one of the great astronomers of the time, William Herschel, royal astronomer to King George III. With royal backing, Herschel was able to construct the finest telescopes in the world, and to embark on a systematic program of observations. Herschel found that the concentration of stars lessened slowly if he looked along the plane of the Milky Way. If he looked perpendicular to the Milky Way, the concentration of stars was lower. Therefore the galaxy was indeed disc-shaped. Toward the end of the eighteenth century, Herschel had a picture of the galaxy rather like two fried eggs put back to back: thicker at the middle, with the stars thinning out toward the edges.

But Herschel and other observers saw more than stars. The sky also contained nebulae, which at first appeared to be just smears of light. In the middle of the nineteenth century, an Irish astronomer, Lord Rosse of Parsonstown, built himself the largest telescope on earth and found that at least some nebulae do have a structure. They appeared to be pinwheels of gas. By 1888, one catalog listed 8,000 nebulae, more than half of them the kind of spiral galaxy that Lord Rosse had found.

No one quite knew what to make of the nebulae. They could be listed as spiral galaxies, as globular clusters, or by their other shapes, but their place in our galaxy was uncertain. What was certain, at least to the great majority of astronomers, was the general scheme of things. The galaxy was pretty much as Herschel had described it —other observers had confirmed his work on the shape of the galaxy—and our sun was close to the galactic center. There was an occasional suggestion that the nebulae might be "island universes," the term used by Kant to describe other galaxies like our own, but the idea did not take root.

THE GREAT GALAXY DEBATE

Enter Harlow Shapley, an American astronomer. A Dane, Ejnar Hertzsprung, had done the work needed to make Miss Leavitt's discovery about Cepheid variables into a usable measuring rod. Shapley used the celestial yardstick to study the distribution of globular clusters. He was able to show that they formed a kind of halo around the Milky Way, and, by combining distance and direction information, he obtained a three-dimensional picture of their distribution. They seemed to be centered on a point lying in the direction of Sagittarius, tens of thousands of light-years distant from our sun. Shapley concluded that this point, not the sun, was the center of the galaxy. In a single stroke, Shapley had not only expanded the dimensions of the galaxy tenfold, he had also demoted the sun from the central position in the universe.

Shapley's estimate of the size of the galaxy later had to be reduced. He did not know that there are clouds of dust and gas that make the light from distant stars look dimmer, and thus he overestimated the distances. His basic picture of the shape of the galaxy, a picture that put the sun near one of its edges, has not been changed. But on one critical point, Shapley was dramatically wrong. He refused to believe that the globular clusters, the spiral galaxies and other nebulae were island universes, galaxies on the same scale as ours scattered through almost unimaginable expanses of space.

Many others shared his views. Battle lines were drawn within the community of American astronomers, pretty much on East-West lines. California was the bastion of progalaxy feeling, while the East Coast was the seat of skepticism. The debate went on through the teens of the century and into the early twenties.

The debate ended on January 1, 1925, at an astronomical meeting which heard a communication from Edwin Hubble, who had been observing nebulae with the 100-inch telescope on Mount Wilson in California. The 100-incher, which was then new and was the biggest on earth at the time, had been able to resolve stars in three nebulae, including the spectacular spiral Andromeda nebula. A study of the periods of Cepheid variables in the nebulae indicated that they were far more distant than any star in the Milky Way galaxy—so distant that the nebulae to which they belonged were clearly separate galaxies. Our galaxy is believed to be some 100,000 light-years across. The galaxies studied by Hubble were at distances of a million light-years or more.

Later, Hubble went on to study even more distant galaxies, hundreds of them at distances of millions of light-years. He concluded that the number of galaxies in the universe is about as great as the number of stars in our galaxy. Astronomers, accustomed to large numbers, casually say that there are 100 billion stars in our galaxy; in an expansive mood, some of them increase the estimate to 200 billion or so. Since our Milky Way galaxy is believed to be a sister to the spiral galaxies seen at great distances, they are believed to have equal numbers of stars. Thus we are talking about perhaps 100 billion galaxies, each with some 100 billion stars. The intellectual odyssey that began with the earth at the center of a small, easily mastered universe has led us to a universe of almost unimaginable extent in which we occupy an inconsequential corner of an inconspicuous island.

Mankind's relatively new knowledge about the vastness of the universe is a key point in almost every aspect of SETI. Are Unidentified Flying Objects visitors from another civilization? How many extraterrestrial civilizations might there be? What is the best strategy to follow as we try to communicate with such civilizations? The size of the universe and the distances both within our galaxy and to the galaxies beyond help define the answers. Our knowledge of the universe also helps answer perhaps the most important question of all: how does life arise? And, in fact, scientific ideas about the origins of life have evolved in step with scientific ideas about the size and structure of the universe. Our picture of the universe is the background against which theories of the origin of life are developed and tested.

The sheer size of the universe seems to say that life on earth is not unique. If the sun is one star of 100 billion in our galaxy, and there are 100 billion galaxies the size of ours, is it possible that life has arisen only once in this universe? Odds of 10,000 billion billion to one are not inviting. It is much more satisfying intellectually to say that earth is just one of the cosmic crowd, a single example of many locations in the universe where life has developed.

But something more than an intellectually satisfying idea is needed. Once people said that Mars was the home of an advanced civilization because that idea was intellectually satisfying. If we are to say that life has arisen many times in the universe, we need a common mechanism by which life can arise. More than that, if we are to say that life elsewhere in the universe is chemically like life on earth, it is necessary to show that the chemistry of earth's lifeforms is not unusual in the universe.

We must go back to the beginning and show how life came to our planet. Then we must show how life could come in the same way to other planets, circling other stars, in our galaxy and in others. To build a case for SETI, science must show how the dead chemicals of an original creation can come to life.

CHAPTER THREE

CHEMISTRY AND LIFE

> If a dirty undergarment is squeezed into the mouth of a vessel containing wheat, within a few days (say 21), a ferment drained from the garments and transformed by the smell of the grain, encrusts the wheat itself with its own skin and turns it into mice. . . . And, what is more remarkable, the mice from corn and undergarments are neither weanlings nor premature but they jump out fully formed.

That recipe for creating life, given in the seventeenth century by the Flemish physician Jan Baptista van Helmont, has been more or less accepted for most of the history of mankind. The Rig-Veda, the oldest Hindu scripture, says that life began from the primary elements. In the West, Anaximander described how the first animals originated from sea slime. Aristotle adopted the same concept in his *Metaphysics*. The idea gave birth to a phrase: spontaneous generation.

It made sense, even in the dawn of the modern scientific era. Anyone with reasonable curiosity could see mice arise from grain and maggots from wheat. Then someone with more than reasonable curiosity, a seventeenth-century Tuscan physician named Francesco Redi, put the matter to a test in 1668. Redi put out eight dishes of meat. Four were covered with cloth and four were not. As the meat rotted, maggots appeared in the four uncovered dishes but not in the four that were covered. The implication was clear: maggots arise in rotting meat because flies lay eggs there, not because of spontaneous generation. Case closed.

FROM "VITAL FORCE" TO "WARM LITTLE POND"

But the believers in spontaneous generation had a riposte. Not meat alone, but meat plus something indefinable in the air was needed for spontaneous generation, they said. The argument went on for a leisurely 200 years. It was brought to a conclusion when money entered the field. The French Academy offered a prize to anyone who could prove or disprove spontaneous generation. Felix Pouchet, a French scientist, entered the lists. So did Louis Pasteur. Naturally, Pasteur won; universal geniuses have a way of doing that. Both he and Pouchet used the same technique: they exposed flasks containing broth to the air. Pouchet said his flasks, which started out sterile, invariably bred microbes. Pasteur said Pouchet got those results because he was sloppy enough to contaminate his flasks. Pasteur finally proved his point by making flasks with long, swanlike necks that allowed air to reach the broth within but trapped any microbes in the air. The broth remained clear of microbial life until the glass necks were broken. The victory went to Pasteur, who said in 1864: "Never will the doctrine of spontaneous generation arise from this mortal blow."

It was a barren victory. As was soon to become clear, spontaneous generation is an acceptable scientific concept. The only nineteenth-century alternative to spontaneous generation was to accept the existence of a "vital force" from some supernatural power that makes living things essentially different from nonliving things. Nineteenth-century chemists were busily undermining the foundation of the vital force concept by making organic molecules in their laboratories. The molecules of living things appeared so complex that it was thought that something special—the vital force—was needed to produce them. Now it became evident that they can be made by a diligent application of knowledge about the carbon atom. If Pasteur had killed spontaneous generation, he had left no acceptable concept of the origin of life in its place.

Svante Arrhenius, the great Swedish chemist, proposed an alternative theory at the beginning of the twentieth century: panspermia. Life on earth had begun because spores or other simple forms of life had drifted here from other planets, other stars. However, panspermia is subject to serious objections. One is difficulty in surviving that any form of life, however simple, would encounter in the cold, radiation-drenched environment of outer space. But even if such a microorganism could survive the interstellar trip, the panspermia concept just pushes the basic dilemma back one notch: How did life arise on that postulated other planet?

The answer was brewing in the writings of another genius of nineteenth-century science, Charles Darwin. He pictured an earth on which living organisms were evolving from simplicity to greater complexity under the influence of purely natural forces. How logical, then, to extend the concept of evolution backward in time and to picture living things evolving from nonliving chemicals. In a now-celebrated letter to a friend, Joseph Dalton Hooker, Darwin wrote in 1871:

> It is often said that all the conditions for the first production of a living organism are now present, which could ever have been present. But if (and oh! what a big if!) we could conceive in some warm little pond, with all sorts of ammonia and phosphoric acid salts, light, heat, electricity, etc., present, that a proteine [sic] compound was chemically formed ready to undergo still more complex changes, at the present day such matter would be instantly devoured or absorbed, which would not have been the case before living creatures were formed.

It was a fascinating, premature idea—premature because nineteenth-century scientific thinking simply could not fit it in. Darwin himself had said as much a few years earlier, in a letter to Hooker: "It is mere rubbish, thinking at present of the origin of life. One might as well think of the origin of matter."

The twentieth century was to think of the origin of both. The pioneer appears to have been a Russian biochemist, Alexander Ivanovich Oparin, who as early as 1924 wrote that "there is no fundamental difference between a living organism and lifeless matter. The complex combination of manifestations and properties so characteristic of life must have arisen in the process of the evolution of matter." Oparin's theory, in which he described "a new colloidal-chemical order" arising "as a result of growth and increased complexity of the molecules" were not published in English until 1938. Well before then, J. B. S. Haldane of Cambridge University, one of the great scientific spirits of the twentieth century, had come up with essentially the same idea.

Both Haldane and Oparin were able to use knowledge about the chemical composition of the universe that had only recently become available. Studying the stars and interstellar gas, astronomers found that the universe consists mostly of hydrogen. A study of the huge outer planets of the solar system, such as Jupiter, indicated the presence of hydrogen-containing compounds such as ammonia and methane. The oxygen that is so plentiful in the earth's atmosphere (which is roughly one-fifth oxygen) was notably lacking. So both Oparin and Haldane started with an earth whose early atmosphere was rich in both compounds of hydrogen and compounds of carbon, mainly carbon dioxide, and that was bathed in ultraviolet light from the sun. The action of ultraviolet light on the early atmosphere produced a vast number of carbohydrates (which, as the name implies, consist of carbon, hydrogen, and oxygen), which accumulated in the primitive oceans until they formed what Haldane called "a hot dilute soup." It is a picture essentially identical to Darwin's "warm little pond."

The difference between Darwin's offhand speculation and the appreciably more detailed theories of Oparin and Haldane was the increase in information about the nature of the chemistry of life and about the nature of the universe. The same sort of scenario was to be played over the following decades: an inspired, informed hypothesis followed by an advance in scientific knowledge which permitted confirmation of the theory.

For example, in 1947, J. D. Bernal, another of the great names in British science at that time, delivered a scientific paper, "The Physical Basis of Life" before the British Physical Society. Bernal took Haldane's "hot dilute soup" one step further. The primeval "soup" was too dilute for the concentrations that would enable the small, primitive molecules to form the large macromolecules needed for reproduction, the most critical function of living organisms. Such macromolecules did not form by themselves in the open sea, Bernal said. He hypothesized that deposits of clay along the seashore helped in the formation of macromolecules: the smaller molecules became attached to the surface of the clay and then joined together. Exactly thirty years later, two scientists working at the National Aeronautics and Space Adminstration's Ames Research Center in California performed an experiment that supported Bernal's hypothesis. The scientists, James Lawless and Nissim Levi, found that a specific kind of nickel-bearing clay attracts the subunits out of which proteins are made. Chains of those subunits form more or less spontaneously on such clay, they said.

THE MECHANISM OF LIFE

Between Bernal's theorizing and the American experiment came one of the greatest achievements of biological science: the discovery of the basic mechanism by which living organisms function and reproduce. Volumes have been written about every step in the road toward the discovery. A summary must suffice here.

Proteins do the work of the cells that make up living organisms. Proteins are made up of subunits called amino acids. Twenty different amino acids are found in the proteins of organisms on earth, in the same way that twenty-six letters are found in the English language. We can form words of any length by stringing letters together, and proteins of any length can be formed by stringing amino acids together. Our hair is protein, our skin is protein, and, more important, the molecules that carry out the basic functions by which living cells eat, excrete and do almost everything else are proteins. To oversimplify grossly, a protein is made of hydrogen, carbon, and phosphorus, with a few other elements, notably sulfur.

Proteins are so important that it was widely believed until the middle of the twentieth century that they are also responsible for storing the genetic information that allows traits to be passed on from one generation to another. To everyone's surprise, that theory was not true. Instead, the genetic material proved to be macromolecules called nucleic acids. A nucleic acid consists of subunits called nucleotides. There are two kinds of nucleic acid: ribonucleic acid, or RNA, so called because its nucleotides contain a sugar called ribose, and deoxyribonucleic acid, or DNA, so called because its nucleotides contain a sugar like ribose, but with one less oxygen atom. RNA is the genetic material for some primitive organisms, but DNA is the genetic material for all reasonably complex organisms, up to and including mankind.

"Genetic material" means this: The DNA molecule consists of two interlocking chains, each containing many thousands or millions of nucleotides. The chains can divide and reproduce themselves, so that each succeeding generation gets an accurate copy of the preceding generation's genetic material. The chains also govern the production of proteins. There are four different nucleotides in the DNA molecule. That is enough to carry the code for the twenty proteins, in the same way that just two symbols, a dot and a dash, can carry the information to pick out any one of the twenty-six letters in the Morse code. A group of three nucleotides codes for a single amino acid. To make a protein, the DNA message is used to produce RNA molecules, which put together the amino acids in the required sequence. Very little of this was known twenty years ago. Today, it is taught by rote to bored high school freshmen.

With this sort of knowledge available, it is possible to carry out an experiment to test the theory of spontaneous generation in a much more sophisticated way than Pasteur did. But the theory that is being tested is far different from that tested by Pasteur. He looked for the spontaneous generation of microbes, which are very complex organisms on the molecular scale. Spontaneous generation today looks for the formation of relatively simple molecules of nucleic acid and protein, under the conditions assumed to have existed at the earliest stage of the earth's formation.

Probably the most influential experiment of this sort was performed in 1953 by Stanley Miller, a student working under Harold Urey, the Nobel laureate in chemistry, at the University of Chicago. Miller created the atmosphere assumed to have existed on the primitive earth—methane, ammonia, hydrogen, and water—in a laboratory flask. Electrodes were placed in the flask, and periodic sparks were sent through them to mimic ultraviolet radiation and lightning. The flask was heated until the water boiled, and the water was allowed to circulate through a closed loop of tubes. Soon the water took on a pink appearance. After a few days, it was deep red. Subjected to chemical analysis, the water was found to contain a number of organic compounds. Significantly, four of the amino acids that are commonly found in protein were among them.

The Urey-Miller experiment was influential not so much because it was the first—a few other experimenters had achieved less advanced results with the same kind of apparatus—but because it showed in convincing fashion that it was possible to produce large quantities of organic compounds starting from what amounted to nothing more than air and water. Much of the work done since then amounts to variations on the theme. The Urey-Miller apparatus is simple enough to be set up in any well-equipped laboratory, and many other scientists soon were at work. For example, Philip H. Abelson, working at the Carnegie Institute in Washington, showed that organic molecules could be produced using a variety of mixtures of gases, as long as oxygen was not the predominant gas. When free hydrogen was present and oxygen was not, the "hot dilute soup" of organic molecules resulted. In more technical terms, the artificial model of the primeval atmosphere had to be "reducing," as opposed to "oxidizing." The earth's atmosphere is now oxidizing—that is, it contains a large amount of oxygen which combines freely with (oxidizes) almost anything it touches. A reducing atmosphere contains little or no oxygen. It does contain such compounds as methane and ammonia, as well as hydrogen, which are known chemically as reducing agents.

Over and over in the past decades, scientists have worked with such a model of a reducing atmosphere to try to produce the basic chemicals of life, with great success. In 1963 Cyril Ponnamperuma and Melvin Calvin of the University of California at Berkeley used an electron beam from a particle accelerator to provide energy to such an atmosphere. German and Soviet scientists have used ultraviolet light. A few experiments have tried to mimic the effects of the shock wave that would be created by a meteorite plunging through the earth's atmosphere. Sometimes different starting materials are used: for example, ammonium formate and hydrogen cyanide, both of which are assumed to have been relatively plentiful on the primitive earth. In the words of Ponnamperuma, "All these different types of experimentation and simulation of prebiological conditions have produced many of the building blocks of life, such as amino acids, purines, pyrimidines, carbohydrates, etc." (Pyrimidines and purines are essential parts of nucleotides.)

However, it could be (and has been) said that these experiments really prove nothing, because they are all aimed at hitting a known target. It can be argued that if you know what kind of molecules you want to produce, you can juggle the elements of the experiment until they produce the expected results.

FINGERPRINTS IN SPACE

There are a number of answers to this argument. The most convincing comes from a study of molecules in space. We are used to thinking of the regions between stars as empty, but clouds of dust and gas are out there. The nature of the molecules in these interstellar clouds can be determined in several ways, all of them based on the emission or absorption of different kinds of energy. Each variety of molecule emits or absorbs specific wavelengths of energy. An emission or absorption spectrum identifies a molecule as surely as a fingerprint identifies a human being. For more than four decades, astronomers have been searching for these fingerprints in space. The first interstellar molecule to be discovered was a two-atom molecule of carbon and nitrogen, called cyanogen, whose chemical symbol is CN, which was found in 1937. Three years later, a two-atom molecule of carbon and hydrogen (CH), methylidyne, was discovered. Today, more than thirty interstellar compounds have been found, from simple two-atom molecules to compounds as complex as ethyl alcohol, the same kind found in martinis.

So we know that some simple molecules related to life are being produced in space. But we know more than that, from the most direct evidence: many complex organic chemicals are being produced in space and are arriving on earth. Their vehicles are meteorites, specifically a kind of meteorite called a carbonaceous chondrite. The organic chemicals that are found in abundance in these meteorites are astoundingly similar to those found in living organisms on earth. The resemblance is so striking that there was a brief time in the 1960s when at least some scientists believed that carbonaceous chondrites were bringing living matter to the earth.

Meteorites have been raining down on earth for as long as can be imagined. Sometime more than a century ago, they began to be collected, and to be named after their landing sites. There is the Orgueil meteorite, which landed in France in 1964, the Murray, which fell in Kentucky in 1950, the Murchison, which fell in Australia in 1969, and so on.

That carbonaceous chondrites contain organic matter has been known for a long time. In 1834, Jons Jakob Berzelius, the celebrated Swedish chemist, reported that he had extracted organic compounds from the Alais meteorite, which fell on France in 1806. Other nineteenth-century scientists made similar studies of other meteorites, with essentially the same results. But the real fun started in the early 1960s, when some scientists began subjecting samples of carbonaceous chondrites to modern methods of analysis. Some of the scientists said they extracted molecules whose existence could be explained only by the action of living things. Other scientists said they had been able to extract real living bacteria from the meteorites.

It made juicy reading, and it started a violent debate. After the arguments and the counterarguments settled down—it took some time, because nothing can be quite as bitter as a polite scientific controversy—a consensus was reached, based on analyses made largely by Ponnamperuma and other scientists at the Ames Research Center, a NASA facility in California. Those analyses were able to explain away a good part of the earlier reports as the result of contamination of the meteorites with terrestrial material. The contamination problem was especially severe for the Orgueil meteorite, which had been on earth for over a century.

But the clinching evidence against a biological origin for the meteorite compounds came from one of the most curious aspects of life on earth—the lopsidedness of amino acids.

Actually, "lopsided" is the wrong word. Amino acids can be right-handed or left-handed. Your right hand and your left hand may seem at first to be identical. But they are not, because they cannot be superimposed on each other. In the same way, your mirror image is not identical to you. And in just the same way, a right-handed amino acid is not identical to a left-handed amino acid.

The amino acids in living organisms on earth are left-handed. No one knows why this is so. The assumption is that life started just once on earth, and that it got off to a left-handed start by pure chance. If this is so—and it is pure hypothesis—we humans and every other living thing on earth can trace our ancestry back to one chance combination in a primeval organic soup.

The amino acids in meteorites are neither entirely right-handed nor entirely left-handed. The meteorites contain equal amounts of right-handed and left-handed amino acids. Ponnamperuma's team first examined the Murchison meteorite, later the Murray and the Orgueil (for the latter, they used a technique which enabled them to distinguish between the amino acids that were present originally and those that were present because of contamination). The results were the same. Not only were the amino acids equally divided between right-handed and left-handed molecules, but some of the amino acids in the meteorites are not found in terrestrial proteins. Ponnamperuma later wrote: "We are therefore led to believe that the amino acids in the Murchison meteorite were produced by an abiotic [nonliving] extraterrestrial process."

Ponnamperuma added a significant comment: "The analysis of this meteorite thus provides us with, perhaps, the first conclusive evidence for the process of chemical evolution occurring elsewhere in the universe." The important word here is "chemical," because it indicates the revival, in a more sophisticated form, of an old idea: Arrhenius's idea of panspermia.

Panspermia has been revived in the same way that spontaneous generation has been revived: in a greatly modified version that fits the facts known to current science. No one conceives of a fully formed life-form arriving by chance from a distant planet, just as no one thinks of living organisms such as maggots arising from river slime. But it is thought possible that many of the essential chemicals for terrestrial life evolved to a large extent in outer space, the "chemical evolution" of which Ponnamperuma spoke. The existence of some of the chemical precursors of life in interstellar space is a starting point for that belief. The presence of organic compounds in meteorites is another indicator. In 1971 Ponnamperuma pointed out that his research team had found the same eighteen amino acids and two pyrimidines, which are essential to nucleotides, in both the Murchison and Murray meteorites. "The findings of this identical complex pattern of amino acids and pyrimidines in two meteorites could mean that this is a basic phase in the chemical process leading to life. This basic sequence in the formation of organic molecules could be determined by the inherent chemical characteristics of the material of our universe," he wrote.

THE SEEDS OF LIFE

Fred Hoyle, a British astronomer who specializes in mind-stretching ideas that sometimes appear to outrage common sense but always stimulate thought, has gone a step further. In collaboration with N. C. Wickramasinghe of University College in Cardiff, Hoyle has proposed that actual living organisms can evolve in interstellar clouds. Given the presence of the molecules known to exist in interstellar clouds as well as grains of dust, Hoyle and Wickramasinghe envision a complex organic chemistry, driven by the radiation that floods outer space, taking place on the surface of the dust grains. The end result, they say, could be a simple living organism, such as a spore, which would be protected from radiation damage by clumps of dust and would feed on the molecules within the cloud.

If that idea is too hard to swallow, how about the presence of tar in outer space? Tar consists of long chains of carbon atoms, with hydrogen atoms attached. In recent years, astronomers have detected a series of interstellar molecules consisting of carbon chains, with some hydrogen and nitrogen. The general chemical formula for such molecules is HC_nN, where H is hydrogen, N is nitrogen, C is carbon and the subscript $_n$ can stand for any number of carbon atoms. Chains of up to seven carbon atoms had been detected by 1977, and a search for more complex molecules was going on. Looking at the sprectrum of radiation absorbed by interstellar dust clouds, Hoyle and Wickramasinghe found "an indication here of a fairly complex, radiation-resistant prebiological polymer associated with interstellar grains . . . . Such prebiological polymers, processed in the interstellar medium, could have played a crucial part in the evolution of terrestrial life."

There are two important implications of chemical evolution in outer space. One is the indication that life elsewhere in the universe could be chemically quite similar to life on earth—that is, built around carbon compounds, and relying heavily on the proteins and nucleic acids that are essential to terrestrial life. Even though Carl Sagan, one of the founders and leaders of SETI, likes to warn against "carbon chauvinism," the conclusion that life elsewhere in the universe is carbon-based is hard to avoid.

Long, complex molecules are necessary for life. Carbon has a unique ability to form such molecules. A carbon atom can form bonds with four other atoms. Because the carbon-carbon bond is unusually strong, carbon atoms can form very long chains that are exceptionally stable at ordinary temperatures. No other element can form such chains. For example, silicon also forms four bonds, but the silicon-silicon bond is not as strong as the carbon-carbon bond, so silicon does not form long, stable chains. Even though silicon is much more abundant on earth than carbon, there is not a trace of silicon-based life on earth. Most of the silicon on our planet has combined with oxygen to form silicon dioxide, the basic ingredient of sand. At our present state of knowledge, a long stretch of the imagination is required to envision anything but carbon-based life forms.

The second implication of the studies of interstellar compounds is that life-forms on other planets are not only carbon-based but that such life-forms could easily be widespread in the universe. Scientists see that the basic building blocks of life on earth form spontaneously in outer space. They see that these building blocks arrive in appreciable quantities on earth. From laboratory experiments, they have worked out a number of alternative pathways by which these building blocks can be put together, in quite ordinary chemical reactions, to form more complex organic compounds. Through a combination of theory and experiment, they have glimpses of the ways in which these organic compounds can combine to produce the self-replicating aggregates of molecules that are the first organisms we can describe as living. And they have a convincing picture of the way in which these first crude organisms could evolve into more efficient and more complex living things.

The picture starts with Haldane's "hot, dilute soup," probably fed by organic material from meteorites. The early earth is a hot planet, whose atmosphere is rich in hydrogen and its compounds. Ultraviolet radiation from the sun penetrates through to the earth's surface easily. Lightning flashes in the sky. With a plentiful supply of raw materials and energy, the building blocks of life, amino acids and nucleotides, are formed out of ammonia, methane and water.

A mechanism is needed to put these building blocks together so as to form macromolecules. Several mechanisms have been mentioned. Leslie Orgel, working at the Salk Institute in La Jolla, California, believes that a molecule called imidazole could have done the job. Imidazole can serve as a catalyst, increasing the rate at which a chemical reaction occurs but not being consumed by that reaction. Orgel's calculations indicate that imidazole could have been formed in the early earth's hot soup by a variety of actions. Imidazole could have put together short chains of amino acids or nucleotides in the absence of water—on a beach, for example.

Imidazole catalysis is not the only possibility. At the Ames Research Center, a research team headed by James Lawless and Nissim Levi, a visiting Israeli, reported in 1977 on the efficiency of ordinary clay as a catalyst.

The Ames researchers found that all metal-containing clays attract amino acids (by definition, all clays contain metals). They also found that one kind of clay, that which contains nickel, attracts exactly those twenty amino acids found in living things on earth. Seven other clays tested at Ames did not have this specific attraction, but they had a quality of equal importance: they destroy amino acids which do not form proteins.

Thus clays could act in two ways to produce the macromolecules of which we are made. Most clays destroy a large number of amino acids, while one specific kind of clay catalyzes the formation of chains of the other amino acids. In a test that simulated the action of the tides—drying an amino acid–clay solution, warming it, wetting it, drying it, and so on for a number of cycles—chains of as many as eight amino acids were formed. Given long periods of time, presumably even longer chains would form.

The Ames researchers found much the same picture in their work with nucleotides and clays. One clay which contains zinc concentrated the nucleotides that make up DNA. Significantly, zinc is known to play an important role in an enzyme called DNA polymerase, which is found in living cells and which links nucleotides to form large DNA molecules.

Those are two possible mechanisms by which macromolecules of nucleic acid and protein could form. Many other mechanisms are possible. The important point is that large, complex molecules can form from simple molecules of the kind that are believed to have been plentiful on the early earth.

Over millions of years, many billions of such macromolecules must have formed on the early earth. Most of them broke down fairly quickly, but the supply was constantly replenished. At last came one of the most important events in the history of the earth. A single molecule somehow managed to put together simpler molecules from its surroundings to produce a copy of itself. The ability to reproduce would give a molecule an evolutionary advantage over all nonreproducing molecules. Soon, copies of that reproducing molecule would begin to become predominant. It is even possible that a single molecule and its descendants filled all the waters of the earth.

Picture many billion descendants of that original molecule being reproduced over many millions of years. The reproduction was not always perfect, especially in an atmosphere in which radiation interfered with chemical processes. New types of molecules were produced by accident. Some of those accidents were more efficient at reproducing than others. The more efficient molecules came to predominate.

As the variety of molecules increased, accidental encounters occurred often. Some molecules that bumped together stuck to each other. The combination was more efficient than two single molecules. Larger and larger aggregates of molecules formed in this way. Gradually, an assemblage resembling a living cell took form. By this time, the supply of food—other molecules—was diminishing because competition was increasing. The next big step in evolution occurred: a primitive assemblage began to make its own food by tapping solar energy through the process called photosynthesis. It was a primitive, inefficient form of photosynthesis, using hydrogen sulfide and giving off sulfate ions, but it worked.

By this time, there may well have been a variety of fairly complex life-forms on earth—complex by comparison with the first macromolecules, that is; they were quite primitive by today's standards. It is believed that the kind of living cell found in higher organisms formed when some primitive life-forms captured and enslaved others. Some of the components of a living cell today still bear the vestiges of a once-independent existence. Just as chain stores swallow mom-and-pop grocery stores, so an enterprising early cell gave itself an evolutionary advantage by swallowing a small but efficient competing cell that provided it with a new capability.

It was well along in the earth's history when another pivotal development—the invention of oxygen respiration—occurred. Anaerobic respiration, using hydrogen compounds, is an inefficient form of metabolism. The first cells that learned how to use water and carbon dioxide to make sugars, releasing oxygen in the process, not only became more efficient but also changed the character of the planet forever. As free oxygen in the atmosphere increased, some of it drifted upward to form an ozone layer that blocked out most of the sun's ultraviolet radiation. (Ozone, a three-atom molecule of oxygen, is formed by ultraviolet radiation and absorbs it effectively.) Free from the peril of ultraviolet radiation, life-forms on earth became more stable. Evolution went on briskly. Here we are.

The rise of more complex life-forms took place over three billions of years. Here we are concerned with the beginnings of it all. From all the laboratory work and astronomical observations, it appears that the capacity for life is not rare, but is woven into the fabric of the universe. The seeds of life—our kind of life, carbon-based and using water—are scattered throughout the universe, waiting only for the right conditions to start germinating. And through their studies of the universe, astronomers also have a reasonably good idea of how often the right conditions occur.

CHAPTER FOUR

THE UNIVERSE AND US

The basic tool of SETI is the radio telescope. While speculations about life on other planets are centuries old, the modern era in which we can talk realistically about communicating with civilizations elsewhere in the galaxy was born only when the radio telescope came of age. More than that: radio telescopy has also helped to give us a picture of the universe that makes possible a reasonable assessment of mankind's chances of finding and communicating with other intelligent life. It was the radio telescope that enabled astronomers to develop their current model of the age, structure, and evolution of the universe. To the old image of the astronomer peering at stars through a huge light-gathering telescope must be added the picture of astronomers reading the jagged graphs that represent the data gathered by bowl-shaped, metallic radio telescopes.

Light and radio waves are two kinds of the same thing: electromagnetic radiation. So are X rays and gamma rays and ultraviolet radiation and infrared radiation. We can measure electromagnetic radiation either by wavelength or by cycles per second. The two measures are directly related: the longer the wavelength, the fewer the cycles per second, and vice versa. That is because all electromagnetic radiation travels at the same speed, which is often called "the speed of light." (That speed is 186,292 miles per second. In the metric system, the speed of light and of all other electromagnetic radiation is a rounder number, roughly 300,000 kilometers per second.)

A wave of visible light is about one hundred-thousandth of an inch long. A radio wave is about half a mile long. You can learn about the number of cycles per second of radio waves by consulting the dial of your radio. "Cycles per second" now are called Hertz, to honor a pioneering scientist. The AM band on your radio is measured in kilohertz, or thousands of cycles per second, while the FM band is measured in megahertz, or millions of cycles per second. If you look at the sky at night, the visible light from stars has a frequency of about ten million billion cycles per second. The distance to those stars is measured in light-years, the distance that light travels in one year—about six trillion miles. Distances of thousands, millions or even billions of light-years are common in astronomy.

TELESCOPY

Since mankind has always looked at visible light from the stars, light telescopy developed early in the modern scientific era. Radio telescopes are developments of the past few decades, and they started through an accidental observation. Karl Jansky, an engineer with Bell Telephone Laboratories, built an antenna in New Jersey in 1931 to study the static that interfered with long-distance telephone calls. He found one kind of radio noise whose source seemed to move with the sun. With some help from an astronomer friend, Jansky found that the radio source seemed to lie toward the center of our galaxy. Soon it became apparent that the radio signals were in fact coming from the center of the galaxy. In addition to visible light, the stars in the galaxy were also emitting radio waves. It was a revolutionary discovery, but when Jansky reported it in 1932, astronomers were mostly indifferent. One amateur named Grote Reber built his own antenna near Chicago and found that the whole Milky Way is a source of radio signals, but professional astronomers did nothing to follow up Jansky's work. Science moved at a more leisurely pace in those days than it does in our era. Now an exciting astronomical observation will be followed up eagerly by many different research teams as soon as it is reported. But Reber did not publish his results until 1940, almost eight years after Jansky's first publication. By then, World War II was upon us, and most purely scientific work was suspended for the duration.

However, the war eventually gave great impetus to scientific research. It was the first war in which scientists were recruited methodically and in large numbers to work on military projects. Anything of potential military use was assured of financial backing, and if the military use was valuable, an almost endless supply of money was available. The atomic bomb was the prime example of this marriage of the military and science. Electronics came close behind, if only because radar had changed the course of the war.

When the war ended, the Cold War began, and the wedding of science and war became even more intimate. All sorts of electronics research boomed ahead, helped greatly by liberal military funding. In the general climate of postwar affluence, even basic research benefited. The large bowl-shaped antennas needed to collect radio signals from the cosmos appeared in every scientifically advanced nation.

The dishes that collect radio signals must be big because radio waves are so long. Big is beautiful in radio telescopy, because bigness improves the quality of reception. There are many movable radio telescopes with dishes that are several hundred feet in diameter, but there is a limit to the size of any single movable antenna. If it is too big, wind and weather will play havoc with it.

There are several ways to get around the size limitation. One way is to build an immovable antenna. The largest single antenna on earth is in Arecibo, Puerto Rico, where a 1,000-foot dish has been built in a natural bowl-shaped geological formation. The builders have done little more than smooth out the circular valley in the green hills of the island, and then lay down acre on acre of precisely engineered metal to gather the radio signals. If it were nothing else, the Arecibo radio telescope would certainly be one of the most beautiful scientific facilities on earth. (See Figure 7.)

The Arecibo observatory has one major disadvantage. The dish cannot be pointed at more than a limited portion of the sky. But its great size gives it many advantages over smaller radio telescopes.

Arecibo also serves as a useful yardstick for SETI. Any number of papers include calculations based on the postulated existence of a radio telescope the size of the Arecibo dish elsewhere in the galaxy.

Radio telescopy has helped settle many major debates about the universe that raged among astronomers a few decades ago. In particular, one discovery by a radio telescope has led to a consensus about the origin and evolution of the universe.

If we opened a book on astronomy and cosmology about fifteen years ago, we would have found explanations of two conflicting theories about the nature of the universe: the big-bang theory and the steady-state theory. Both of them were attempts to explain the observation that the universe is expanding.

"BIG BANG" VS. "STEADY STATE"

Edwin Hubble discovered the expansion of the universe in the 1920s, using a light telescope. He found that the light emitted by distant galaxies is not the wavelength that it should be. The light is shifted toward the red—or longer wavelength—end of the visible spectrum, because the galaxies are moving away from us. We get the same effect on earth when we hear an automobile horn appear to wail as the auto speeds away; the sound waves of the horn are stretched out because the auto is moving away. In the universe, the "red shift" of a given galaxy is directly related to its distance from earth: greater distance means a larger red shift, and hence a greater rate of speed.

Picture dots on a balloon. If the balloon is inflated, the dots will move apart. From the vantage point of any single dot, it will appear that all the others are flying away, and that the most distant dots are moving away the fastest. So it is with the universe, except that the dots are galaxies.

One explanation of the expanding universe took a motion picture of the speeding galaxies, ran the film backward and arrived at a moment when all the mass in the universe was in one "superatom." Some unexplained explosion—a "big bang"—sent the matter streaming out, this theory said. An alternative theory said that the universe has always looked like this. The steady state of the universe

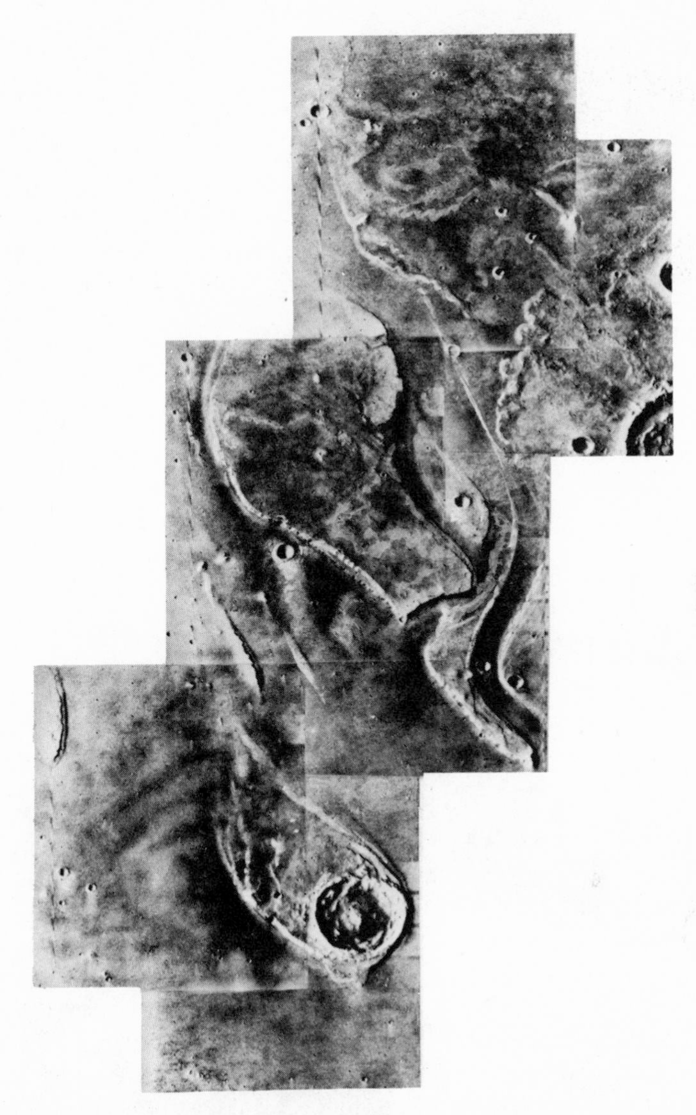

FIGURE 1 The northeast portion of the Chryse region of Mars, as seen from the Viking I orbiter. The meandering, intertwining channels are believed to have been cut by running water.

FIGURE 2 The first picture taken on the surface of Mars. It shows one of the three footpads of the Viking I lander and a patch of Martian soil.

FIGURE 3 The first panoramic view of the Martian surface, as seen from the Viking I lander.

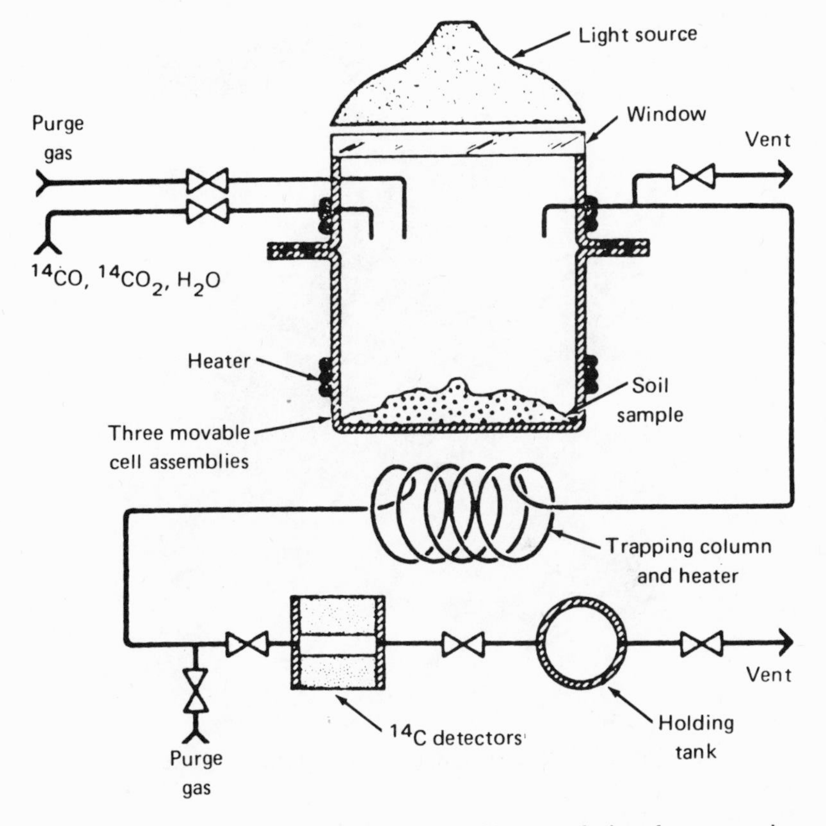

FIGURE 4 A schematic drawing of the Viking pyrolytic release experiment.

FIGURE 5 A trench gouged in the Martian surface by Viking.

FIGURE 6 A proposed surface rover for a future Mars mission. Running on tracked wheels, the rover would carry more than 220 pounds of scientific instruments and two stereo cameras mounted on the columns in front.

FIGURE 7 The radio telescope at Arecibo, Puerto Rico. The antenna, built into a mountain valley, has a diameter of 1,000 feet. The Arecibo Observatory is part of the National Astronomy and Ionosphere Center, operated by Cornell University under contract with the National Science Foundation.

FIGURE 8 The planets Mercury, Venus, Mars, Saturn, Uranus, and Jupiter.

FIGURE 9 Mars, as photographed from 348,000 miles by Viking I on June 17, 1976.

is maintained by the constant creation of new matter, according to this theory.

The big-bang theory is accepted now, because radio astronomers can detect a faint background noise that pervades the entire universe. This noise is the remnant of the big bang, the explosion of the superatom that started the expansion of the universe some fifteen to twenty billion years ago.

Starting with the big bang and using a combination of theory and observation, astronomers have developed a picture of the evolution of the universe. Many details are uncertain, but the picture is reasonably complete, from the largest scale—the origin and history of galaxies—to the smallest. Among the smallest details in this cosmological picture is the formation of planets such as the earth. Another detail is the origin of life on such planets. On the cosmological scale, a planet such as the earth is no more than a pebble on a very large beach, and life on the planet is a thin film of scum on the pebble. But the theory that is accepted by many astronomers makes both the formation of planets and the existence of life on them an integral part of the history of the universe.

After the big bang occurred, the universe consisted mostly of hydrogen, with some helium thrown in. Hydrogen is the simplest element; its nucleus consists of a single particle called a proton. The helium nucleus consists of two protons and two neutrons.

As this gas streamed out from the big bang, eddies formed. The expansion of the universe was constant, but matter was not distributed evenly through space. Instead, it tended to clump together. On the largest scale, those clumps are the galaxies. They are truly island universes. Everywhere that astronomers look in the universe, they find galaxies floating in the sea of space. Galaxies come in a variety of shapes, from globular clusters, which are spherical, to flat discs of billions of stars, with spiral arms as if they were pinwheels spinning in space. The Milky Way galaxy is such a disc.

Galaxies are spinning. They must, or they would collapse into a single mass as a result of the gravitational attraction of the matter they contain. This rotation is important for the formation of stars. Just as huge eddies lead to the formation of galaxies, eddies on a smaller scale lead to the formation of stars within galaxies. The two processes are essentially identical, except for scale. Just as galaxies always form in clusters, stars always form in clusters.

Envision an eddy of gas and dust, a cloud just a few degrees above absolute zero. The cloud begins to collapse because of the gravitational attraction of its matter. Gravitational energy produces heat and radiation. As the collapse continues and the density of the cloud increases, more and more of this energy is trapped within the cloud. Its temperature rises. At one time or another, the amount of pressure created by the radiation almost equals the gravitational attraction. The cloud is what we call a protostar. It is about the size of our solar system and is collapsing steadily. Bart Bok of the University of Arizona has discovered small, dark bodies that have all the characteristics of protostars. These Bok bodies are believed to be stars in the process of formation. Other astronomers have detected what seem to be stars in the earliest stages of their birth.

Both gravitational collapse and temperature rise continue until the center of the cloud becomes so hot and dense that the hydrogen nuclei are squeezed together. Very great pressure and high temperatures are needed for the fusion of hydrogen nuclei, but it does occur in the interior of stars. The fusion of four hydrogen nuclei forms one helium nucleus and releases a great deal of energy. On earth, we are familiar with fusion in hydrogen bombs, which have given mankind the ability to destroy itself. If we do wipe out life on earth by using hydrogen bombs, we deserve no credit for originality. We have simply borrowed from the stars.

STARS ARE NOT CREATED EQUAL

Stars are not created equal. Some are brighter than others. Some burn low, with a dim, red glow, while others burn with a white intensity. For something close to a century, astronomers have been classifying stars by their brightness and temperature. One of the methods of classification uses letters, in a quite illogical sequence. Starting with the hottest stars and moving toward the cooler, the types are listed as:

O, B, A, F, G, K, M,

which can be remembered by the classic mnemonic, "Oh be a fine girl, kiss me."

It is possible to plot a graph of stars which relates their temperature and their brightness. Such a graph is called a Hertzsprung-Russell diagram, after the two astronomers who developed it. If brightness is graphed from top to bottom and temperature is graphed from left to right, most stars fall in a broad, wavy band that runs from the upper left to the lower right. This band is called the "main sequence." Our sun is pretty close to the center of it.

The sun has been on the main sequence for five billion years, and will be on it for another five billion years. The sun is a G star, pretty much a stellar mediocrity. There are many stars like the sun on the main sequence, a fact that is very important to SETI. But the sun is also a particular kind of main-sequence star, which is also important in the matter of life in the universe—specifically in the matter of life in the solar system.

The mass of a star determines both its place and the time of its stay on the main sequence. Stars with low mass are cooler and thus less bright; they are toward the lower part of the main sequence, and they will spend a long time there. Because the temperatures and pressures inside them are relatively low (by stellar standards), these small stars will burn their hydrogen fuel at a moderate rate for a long time. At the upper end of the main sequence are the most massive, hottest, brightest stars. They will be gone in a twinkling—a million years or so. There are size limits at both ends of the main sequence: nothing with a mass less than about one-twelfth that of the sun can achieve fusion, and stars with masses 100 times or more that of the sun probably do not exist.

Stars die, often explosively. Stellar death is a fascinating and complex process, but we are interested in only a few major points. Death occurs because the star uses up its fuel. Hydrogen can be fused into helium for just so long, until the hydrogen is almost exhausted. Helium nuclei can be fused to form heavier nuclei, and those nuclei can be fused to form even heavier nuclei; but there comes a time when the energy output of the star cannot balance the gravitational pull of its matter. If the star is massive, it explodes, sending a good fraction of its material out into space.

That material joins the clouds of interstellar gas and dust, and some of it eventually is incorporated into new stars. Unlike the stars we have described until now, the new stars start off with more than just hydrogen and helium. They contain heavier elements—carbon, oxygen, nitrogen, sulfur, and phosphorus, among others—formed in fusion reactions in the old stars.

Stars that contain these heavier elements are called second-generation stars. They get the heavier elements from first-generation stars, which contain only hydrogen and helium. The sun is a second-generation star. When the sun set up shop some five billion years ago, the universe had been in existence at least two, and perhaps as much as three times longer than that. Time is an element that cannot be forgotten in SETI, and we will return to it.

Spin is important to stars, too. As an eddying cloud of gas and dust begins the contraction that leads to formation of a star, it is spinning. The more the cloud contracts, the faster it spins. Before long, the cloud looks something like a football: a dense central bulge, which will become the star, surrounded by a halo of dust and gas, which will become planets.

Dwell for a moment on that last sentence. It assumes quite matter-of-factly that the formation of planets is a routine affair in the history of a star. Yet less than half a century ago, Sir James Jeans, one of the great astronomers of the century, proposed a theory that would have made the creation of our solar system an unusual—perhaps unique—event, the product of a near-collision between our star and another. It was a theory that many scientists could accept when we lived in a smaller, more manageable universe where the earth could be regarded as something out of the ordinary. Today, when we know that the earth is just a dust mote in space, we have a theory in which the earth and the entire solar system are all too ordinary.

Back to the spinning protostar. The central bulge is surrounded by a disc of dust and gas, the solar nebula. Grains of dust in the nebula begin to stick together, forming larger particles. As fusion reactions switch on and the sun begins to radiate heat, lighter gas is driven toward the edge of the disc, while heavier material stays toward the center. Close to the sun, the dust particles accumulate to form pebbles, rocks, boulders—eventually, planetesimals—small planets—that pull in more matter until full-sized planets form. Out away from the sun, the planets that form are both bigger and lighter, consisting mostly of the gases such as hydrogen and helium that were in the solar nebula. In our solar system, the four inner planets—Mercury, Venus, Earth and Mars—are rocky, while the planets farther from the sun—Jupiter, Saturn, Uranus and Neptune—are gaseous. (Pluto, the ninth and outermost planet, is something of a mystery.) Meteorites are believed to have formed at the same time as the planets, and thus they give us a sample of the earliest material of the solar system.

There is reason to believe that this distribution of matter is typical of other planetary systems even though there is no ironclad evidence that other solar systems exist. Stephen Dole, an American astronomer, has run computer simulations of the formation of planetary systems from a solar nebula. Starting with a typical solar nebula, Dole's computer program prints out a series of planetary systems remarkably like ours—rocky planets close to the sun, gaseous planets farther out. The sizes and distributions of the planets vary somewhat, but the resemblance between the computer-generated planetary systems and the solar system is unmistakable.

Dole's computer program even generated situations in which one of the bodies orbiting the protostar was large enough to become a star itself. Binary (two-star) systems are common; in fact, most of the stars we can see from earth are in multiple-star systems. It appears that the difference between a star and a planet is only one of mass. If Jupiter, the largest planet in the solar system, were about 80 times more massive, it would be large enough to generate fusion reactions, and the sun would be part of a binary system.

Direct evidence for the existence of planets around other stars is controversial, but spin gives indirect evidence that planetary systems are common. The rate of rotation of a star should be in direct relationship to its mass: the more massive the star, the faster it should spin. This relationship holds, but only up to a point—a most interesting point for SETI. Hot, bright stars—the more massive ones, —do spin rapidly. They can take only a few hours for a complete rotation. By this standard, our sun should complete a rotation in less than a day. In fact, the sun revolves only once every 25 days. And the sun is not unusual. Stars that are about the size of the sun spin much more slowly than they should. Astronomers have found an abrupt change in rotational speed in stars whose mass is about that of the sun. Something has to slow the rotation of these stars, and planetary systems seem to be the logical explanation.

But there is a pesky problem to which astronomers did not have a solution until recently: how is spin (or, to use the formal term, angular momentum) transferred from a star to its planets? One reason why Jeans and other astronomers postulated a collision between stars to explain the existence of the planets was the lack of a theory to explain why the planets had a much greater share of the angular momentum of the solar system than they should. Today it is believed that the angular momentum was transferred from the protostar to the planets by means of an intense magnetic field. According to this theory, the protosun created a magnetic field in the gas halo that surrounds it, and the effects of this magnetic field transferred angular momentum from the protosun to the planets-to-be.

One of the most significant points of this theory for SETI is that it paints an extremely optimistic picture of intelligent life in the universe. Not only on theoretical grounds but also from observation, we know that life emerged on earth very early in the history of the planet. On the principle of mediocrity, we can assume that life is an early arrival on earthlike planets in other solar systems. However, again from the principle of mediocrity, we must assume that *intelligent* life makes a relatively late appearance. On earth, the interval between the production of the first life-forms and the presence of intelligent life is much more than three billion years.

Therefore, a giant star that burns out its fusion fuel in a few hundred million years, or even in a billion years, is not likely to have a planetary system that could harbor intelligent life. Even if an extremely massive star does have earthlike planets, there simply will not be enough time for intelligent life to evolve on those planets before the star dies. Only longer-lived stars, those that spend several billions of years on the main sequence, are likely candidates for intelligent life. And, as we have just seen, the rotational evidence indicates that it is just such stars, those with masses about that of the sun, that seem to have planetary systems. If Dole's computer simulations are an accurate reflection of reality, many of those planetary systems include earthlike planets.

But even if the galaxy abounds with long-lived stars that have planetary systems, and even if many of those planetary systems include planets like the earth, how many of those postulated planets are indeed capable of supporting life. Su-Shu Huang, a Chinese-born American astronomer, has tried to answer that question by developing a theory of a "zone of life," a region in the vicinity of a star where conditions are right for the development of life.

Temperature is a major factor in the size of a star's life zone. We assume that liquid water is needed if a planet is to support life. That assumption can be attacked as being too narrow, but most of the evidence that is now available indicates that liquid water is essential for living organisms. Therefore, the life zone of a star is the region where the temperature is high enough to keep water liquid but not high enough to turn water into steam. If we look at our own planetary system, we find an excellent illustration of this principle. Venus is an earthlike planet whose temperatures are too high for liquid water to exist. Mars is an earthlike planet whose temperatures are too low for water to become liquid. The earth nestles in the sun's life zone, and it supports an abundance of life.

When we consider the life zones of other stars, time as well as temperature must be taken into account. It would appear at first glance that large, hot stars have larger zones of life than small, cool planets. The large star radiates a lot of energy, and so the region in which the temperatures are right for liquid water are also large. By the same reasoning, the life zone around a star that is small and therefore cool will be narrow. Large stars would therefore have more habitable planets within their life zones than small stars.

But when time is added to the equation, the balance swings in favor of smaller, cooler stars, those whose mass is about that of the sun. It is true that a star of this size has a smaller zone of life than a massive star, which means that there is a smaller chance that any planet will be in the zone of life. But if a smaller star—again, about the size of the sun—does have a planet in its life zone, there will be time aplenty for intelligent life to evolve on that planet. A small star spends billions of years on the main sequence. By contrast, a more massive star will probably burn itself out before intelligent life can evolve on any of its planets.

Of course, much of this theorizing is sheer speculation. Life could evolve much faster on a planet orbiting a hot star than it does on a planet such as earth. It is known that the rate at which chemical reactions take place speeds up as the temperature rises. The tempo of life, death, and evolution could be rapid on a planet that receives intense radiation from a massive star. We can visualize fast-evolving life-forms that live and die in a day, as some summer insects on earth do. We can picture sentient beings moving and thinking at a tempo that makes life on earth look like a slow-motion film.

But the principle of mediocrity steps between us and this kind of speculation. In the search for extraterrestrial life, it is important to think not of what is possible but of what is probable. High-speed evolution on an unearthly planet orbiting a blue-white massive star is not probable. In addition, the evidence of stellar rotation indicates that planetary systems are most likely to form around smaller stars. The most sensible conclusion is that other intelligent species in the universe, if they exist, live on planets like the earth that orbit stars like the sun.

What about binary stars? The standard assumption has been that two-star systems are not suitable for habitable planets because any planets in a binary system would have complex orbits which would move them from freezing cold to boiling heat. In other words, planets in a binary system would move in and out of the zone of life —not a comfortable situation for living organisms. However, recently, Robert S. Harrington of the U. S. Naval Observatory in Washington has performed some calculations that make the picture appear more optimistic. Harrington found that the orbits of planets around two-star systems could be stable enough to keep planets in the life zone. For example, Harrington calculated that if the solar system was altered by replacing Jupiter with a star as big as the sun, only Mars would be thrown into an unstable orbit. Under the proper conditions, he found, the normal habitable distances for earthlike planets are in the zone of stability for at least some binary stars.

In the midst of all this theorizing, it would be nice to have a solid fact or two about other planetary systems—for example, a straightforward astronomical observation of a planet orbiting another star. Unhappily, direct evidence that other planets exist is extremely hard to come by. Unless it is very large, a planet is a negligible object on the cosmic scale. However, it is at least theoretically possible to detect the existence of a planet by its effect on the motion of its star. If the planet is massive enough, its gravitational pull will be great enough to tug the star slightly in its direction during its orbit. A distant observer will not be able to see the planet, but the wobble in the star's motion that is induced by the gravitational pull of the planet will be visible.

Seen from the earth, stars appear to be moving through space. This apparent motion is due to the spin of the Milky Way galaxy. As the galactic disc whirls, the sun goes around and around, as if on a cosmic carousel. Anyone who has ridden a carousel knows that it is possible to create the illusion that everything else is moving. Seen from the earth, stars should move along smooth paths. A wobble in the path means that a star is being tugged by the gravitational pull of another body. As astronomer George Gatewood of Allegheny Observatory in Pittsburgh explains: "Imagine two people dancing a polka. As they whirl about to the music, each person—pulled by the other—moves in a series of circles across the floor. Now imagine that one of the dancers is invisible; we can still tell he is there by the motion of his partner."

There are strict limits on this kind of observation. Either the star must be small or its companion must be large. And a star must be relatively near the earth for a wobble to be detected by our astronomers.

BARNARD'S STAR

The best illustration of the difficulties in the observations of such stellar motions—the field is called astrometry—is the story of Barnard's star, so called because it was discovered in 1916 by Edward E. Barnard of the Yerkes Observatory. Barnard's star is both small and nearby; at a distance of 5.9 light-years, it is the second nearest star to earth. For more than three decades, an unusually detailed record has been made of the motion of Barnard's star, a record that is written on some 10,000 photographic plates made at the Sproul Observatory at Swarthmore College in Pennsylvania under the leadership of astronomer Peter van de Kamp. In 1964 van de Kamp announced that the observations indicated that Barnard's star has a planet that is about half again as large as Jupiter, the largest planet in the solar system.

The announcement produced a sensation, because of the nearness of Barnard's star. A few years later, van de Kamp said a reanalysis of his data indicated that Barnard's star actually has two planets, one of them 1.1 times the mass of Jupiter, the other 80 percent as massive as Jupiter. For a decade, the planet (or planets) of Barnard's star were accepted as the real thing.

Then, in 1976, Gatewood said that van de Kamp was wrong. On the basis of observations made at the Allegheny Observatory and at the Van Vleck Observatory in Connecticut, Gatewood said that the evidence given by van de Kamp for the existence of the two planets around Barnard's star was erroneous. But Gatewood added that his observations indicated the existence of a planetary system around Barnard's star—two planets, the size of Jupiter and Saturn. Van de Kamp checked his data again and said that he was right in the first place. At the moment, the issue is still undecided, although the volume of the Sproul observations makes many astronomers lean toward believing van de Kamp.

Somewhat the same story is being played out about Epsilon Eridani, a star whose mass is about the same as that of the sun. Van de Kamp says that there is evidence indicating that Epsilon Eridani is orbited by a planet with six times the mass of Jupiter. Gatewood's observations cast doubt on that possibility.

The debate about astrometric evidence for other planetary systems is likely to go on for some time. As Gatewood pointed out in 1976, astrometry required periods of observation "similar to the orbital periods of the objects sought." Those orbital periods are quite long—12 and 26 years, respectively, for the planets which van de Kamp believes have been detected around Barnard's star. Astrometrists think in decades.

And they do not think about making contact with civilizations on any planets that they might detect. Such planets are just too big to support the kind of life that we are familiar with. All that astrometry can offer SETI is the possibility that the cosmos is swarming with planetary systems. For example, Helmut Abt of the Kitt Peak National Observatory in Arizona studied 123 stars resembling the sun and concluded that almost all of them had companions. Most of those companions were other stars. By extrapolating from the patterns he observed to smaller masses than he could detect, Abt concluded that as many as one-third of the stars could be orbited by bodies smaller than stars. However, those bodies could not quite be described as planets. Although not large enough to ignite the fusion reactions that would make them stars, they are massive enough to be in gravitational collapse, generating enough energy to remain starlike in appearance for many millions of years.

If Abt's figures are extrapolated down further, the uncertainties become unacceptably large. It is even impossible to use his observations to estimate the number of Jupiter-sized planets orbiting other stars. Earth-sized planets are so small on this scale that they are undetectable.

But there is one place where earth-sized planets are detectable: the solar system. The planets orbiting the sun may be a small sample, but they are all we have. One of them we can examine directly—a convenience, since it happens to be the only one known to support life. The others can be investigated by a variety of methods. And they have a lot to tell us.

CHAPTER FIVE

ONCE AROUND THE SUN

Mercury we can write off at once: too darned hot. The closest planet to the sun is not at all a likely candidate for life. Mercury is about the size of the earth's moon, and it has an eccentric elliptical orbit that brings it within 29 million miles of the sun at its closest approach and nearly 44 million miles away at its most distant point. Mercury rotates slowly, making a complete turn only once every 59 earth days. Temperature measurements from earth indicate that the side of Mercury that faces the sun is at about 1,800° F., while the dark side is below zero. With these temperature extremes and the virtual absence of an atmosphere, Mercury is about as dead as a planet can be. (See Figure 8.)

Venus is a different story, and a fascinating one. The fascination commences with the strong resemblance of Venus to earth. Venus is about the same size as the earth and has about the same density, indicating that it is made of much the same material. Venus even receives about the same amount of sunlight as the earth does; although Venus is much closer to the sun than the earth is, the thick blanket of clouds in which Venus is swathed reflects a good deal of sunlight into space. In the years when belief in life on Mars captured the popular imagination, scientists such as Svante Arrhenius speculated freely about lush, dripping forests on the surface of Venus, rather like the tropics on earth.

HELL ON VENUS

Until the middle of the twentieth century, not much was known about the surface of Venus. Now we know a lot—too much, in fact, for those who envision a lush Venus. The disillusionment began in the 1950s with earth-based measurements which indicated that the surface temperature of Venus is at least 500° F., a reading which meant that Venus would also have an atmosphere at least 10 times more dense than earth. Those figures seemed surprisingly high at the time. Now they seem conservative.

A series of spacecraft launched by the United States and the Soviet Union have give detailed information about the surface of Venus. The United States concentrated on fly-bys, while the Soviets have sent many probes into the atmosphere and toward the surface of Venus. The U.S. missions confirmed the high temperature and density of the atmosphere. After a long series of failures, the Soviets succeeded in sending a probe into the atmosphere in 1967. It was burned and crushed to death in the hot, dense atmosphere, many miles above the surface. The Soviets made tougher, better-armored spacecraft; they too, were destroyed by the hostile conditions. In 1970 a Soviet craft, Venus 7, stayed alive on the surface for a few minutes, reporting a temperature of 900° F. and a pressure 90 times greater than earth.

Most planetary scientists are certain that Venus does not support life. Conditions on Venus certainly seem as hostile as could be imagined for terrestrial organisms. The atmosphere consists almost entirely of carbon dioxide, with little or no water. The yellow clouds, many miles thick, in which the planet is wrapped, are believed to be made up of sulfuric acid. And the intense heat and incredibly dense atmosphere make the planet literally hellish.

Some scientists have not give up hope for life on Venus. Bruce Murray, a planetary scientist who heads the Jet Propulsion Laboratory in Pasadena, California, believes that there could be organisms living in the atmosphere of Venus, at an altitude where temperatures are more moderate than they are on the surface. Such organisms could feed on the sulfur compounds in the clouds of Venus, Murray says, spending their entire existence aloft. Murray's idea is imaginative, but the consensus is that life is improbable on Venus.

There have been proposals for making Venus more receptive to terrestrial life, by a major long-term effort in planetary engineering. When the hostile nature of its atmosphere became known, there was some talk of seeding Venus with organisms that could make the planet habitable. A spacecraft would scatter blue-green algae across the upper atmosphere. These microbes would reproduce, spread across the planet, and break down the carbon dioxide of the atmosphere to liberate oxygen. A healthy rain would fall, and Venus would become the tropical paradise of earlier dreams. Those hopes now have been largely abandoned, because of the low probability of an organism surviving the sulfuric-acid attack and the absence of water. If planetary engineering ever is attempted on Venus, some way of importing vast amounts of water must be found, and a number of other problems must be solved. It will not happen soon.

But fascination with Venus still continues, because of the resemblance between Venus and earth. Why should one planet be so hospitable to life and the other be so hostile?

Part of the answer comes from the chemistry of the solar nebula from which the planets formed. John S. Lewis, a geochemist at the Massachusetts Institute of Technology, theorizes that different chemicals are present at different distances from the protostar in a solar nebula. The region in which earth formed had water-containing silicates, Lewis said, while the region where Venus formed did not. Earth got water; Venus got none, and that has made an enormous difference.

Aside from water, there is not much potential difference between the atmosphere of earth and that of Venus. However, the prevailing theory is that planetary evolution followed a far different scenario on Venus. Because it is nearer the sun, Venus gets 1.75 times more solar radiation than the earth. Much of the radiation is reflected away by the thick cloud blanket. But the radiation that enters the atmosphere is trapped in the clouds and in the carbon dioxide atmosphere. On earth, living creatures lock up much of the carbon in such things as seashells. On Venus, the absence of life allows sunlight to liberate carbon from rocks. The carbon dioxide which is formed allows the shorter rays of solar radiation to enter but traps the longer waves of radiation that are reflected by the soil. More carbon dioxide traps more heat, and the heat liberates more carbon to form carbon dioxide. The result is a runaway "greenhouse effect," so called by analogy with a heat-trapping greenhouse on earth (although a greenhouse actually traps heat by holding hot air beneath its glass).

Could a runaway greenhouse effect happen on earth? Some scientists nervously suspect that it could. All the ingredients are present. If all the carbon in earth's limestone and seashells were liberated, the density of the atmosphere would increase seventyfold. If the temperature rose enough, water would be broken down into hydrogen and oxygen. The hydrogen would escape from the earth's gravitational pull, and the oxygen would combine with surface minerals. One nightmare is that mankind might manage to upset the ecological balance of earth enough to produce another Venus.

LIFE ON EARTH

But earth teems with life, and it appears to have been well stocked with living creatures for billions of years. One of the more encouraging developments for SETI in recent years has been an accumulation of evidence showing that life established itself on this planet with remarkable speed and has flourished under almost every conceivable terrestrial condition.

A scientist who has done a great deal to push back the limits of life on earth is Elso Barghoorn, a paleontologist at Harvard University. Working with minerologist Stanley Tyler in the 1950s and 1960s, Barghoorn showed that there are fossil microbes in a Canadian rock formation called the Gunflint, which is two billion years old. In 1977 Barghoorn pushed the frontier back considerably further, by showing that a 3.5-billion-year-old rock formation in southern Africa contains fossil microbes that appear to be primitive forms of algae.

Consider that the earth formed from the solar nebula about 4.5

billion years ago, that millions of years were needed for the evolution of the first molecules of life, that more time was needed for those molecules to form themselves into a primitive cell. One billion years for the appearance of such cells is an impressive performance.

Primitive cells could have appeared even earlier. Late in 1977, Carl R. Woese and some other microbiologists at the University of Illinois said that they had identified a small family of microbes as being distinctly different from all other organisms on earth—different enough to be descendants of the first microbes to have evolved on earth.

Woese's microbes are anaerobic: they do not require oxygen to live. Most microbes and all higher life-forms are aerobic: they require oxygen. The family of organisms studied by Woese has members that are found only in oxygen-free environments: the intestines of ruminant animals, rotting sewage, stagnant marshes, hot springs in Yellowstone National Park. The microbes "breathe" hydrogen and carbon dioxide and excrete methane. Until Woese made his report, these microbes were classified as bacteria. But a long and detailed study of the genetic material of the organisms convinced Woese that they are different enough from other bacteria to constitute a separate form of life. If Woese is right, the standard textbook division of all terrestrial organisms into prokaryotes—simple organisms whose genetic material is not gathered into a cell nucleus—and eukaryotes—more complex organisms which have cell nuclei—must be changed. Now the classification must include another kind of prokaryote, the organisms that Woese tentatively called "methanogens," as a third life form.

Woese believes that the ancestors of the methanogens evolved about four billion years ago, well before the algae whose fossils were discovered by Barghoorn. If Woese is right, then organized life on earth got off to a running start, not long after the planet began to cool from its fiery beginning. The methanogens breathed in the hydrogen compounds and carbon dioxide of the primitive atmosphere and breathed out methane that entered the evolutionary cycle. As Ponnamperuma said when he heard of Woese's report, "The idea fits in beautifully with the idea that life began in nonoxygenic conditions." Score one for the SETI concept of a universe where life is a common occurrence.

Against that possible victory must be marked the apparent failure to find life on Mars in the Viking mission. However, the hope that there might be some kind of life on Mars is not dead. There is the plain evidence from Viking orbiter pictures that water once flowed in abundance on the Martian surface. Some planetary scientists, like Carl Sagan, have proposed that Mars goes through climatic cycles, in which conditions change from being hostile to life to being friendly and back again. In the past few years, geologists have produced convincing evidence that the earth goes through climatic cycles. The driving force behind these cycles is apparently small irregularities in the earth's orbit and spin. These irregularities affect the amount of solar energy received by the earth enough to cause such large-scale effects as the ice ages.

MARS: THE UNTOLD STORY

But most planetary scientists now believe that the flow of water on Mars was a once-only phenomenon, something that happened relatively early in the history of the planet. The key to this belief is the relatively small size of Mars, which has half the diameter and only 38 percent of the gravitational attraction of earth. Assume that Mars, at birth, had the same content as earth of what scientists call "volatiles"—gases and liquids. Earth had the gravitational pull to hold on to an appreciable fraction of its volatiles; Mars did not. The thinness of the Martian atmosphere does nothing to mitigate the cold of Mars. There is plenty of water on Mars, scientists calculate; the polar caps, under a veneer of dry ice—frozen carbon dioxide—are made of water, and the Martian soil is believed to contain enough frozen water to cover the planet hundreds of feet deep, if it were melted. But water does not melt on Mars. It sublimes—that is, goes directly from ice to vapor—which is not much help to living organisms.

Maybe there is life on Mars, even given all this. Perhaps organisms evolved in the early, lush days, and have burrowed in for survival. We know of microbes on earth that can survive under conditions about as severe as those found on Mars. The Viking spacecraft literally scratched the surface of the planet. Future missions will dig deeper. An optimist such as Carl Sagan can rattle off several scenarios which end with the survival of Martian microbes. The story of Mars is not all told yet. (See Figure 9.)

Assume the worst: that Mars is indeed lifeless. The first reaction is that of bitter disappointment for SETI, because the life zone around the sun appears to shrink. But a closer look allows a more cheerful interpretation. Edward Anders of the University of Chicago and Tobias Owen of the State University of New York at Stony Brook took a closer look after the Viking mission. They concluded, in a scientific paper, that "Mars was poor in volatiles from the start" and fell behind earth not only by losing lighter gases but also because more volatiles were—and are—trapped in the Martian soil. But they had this to add:

> If it turns out that Mars is completely devoid of life, this does not mean that the zones around stars in which habitable planets can exist are much narrower than has been thought. Suppose Mars had been a larger planet—the size of Earth or Venus—and therefore had accumulated a thicker veneer and had also developed global tectonic activity on the scale exhibited by earth. A much larger volatile reservoir would now be available, there would be repeated opportunities for tapping that reservoir, and the increased gravitational field would limit escape from the upper atmosphere. Such a planet could have produced and maintained a much thicker atmosphere, which could have permitted at least an intermittently clement climate to exist. How different would such a planet be from the present Mars? Could a warm, stable climate be maintained? It seems conceivable that an increase in the size of Mars might have compensated for its greater distance from the sun and that the life zone around our star would have been enlarged accordingly.

In other words, a different role of the cosmic dice might have made this a solar system with two inhabited planets. And since the dice have been rolled billions of times as stars come into being, many planetary systems with more than one habitable planet could exist.

With their habit of thinking big, scientists in SETI have raised the possibility that our solar system may yet be made to have two habitable planets, even if mankind must take a hand. In 1976 a committee of exobiologists working out of the Ames Research Center in California issued a report, "On the Habitability of Mars: An Approach to Planetary Synthesis," which examined what mankind might do in the way of planetary engineering on Mars.

There is nothing small about the way these scientists think. Is Mars too cold and its atmosphere too small to support life? Then spread sand or dust on the polar ice caps to increase the amount of sunlight they absorb, vaporizing the caps to thicken the atmosphere, and raise the temperature. Are terrestrial microbes unsuited for the rigors of Mars? Then create new microbes, using genetic engineering. ("Indeed, in principle, the entire gene pool of the earth might be available for the construction of an ideally adapted oxygen-producing photosynthetic Martian organism," the report said.) The time required for all this? Perhaps 100,000 years or more, although the transformation might be accomplished in as little as 10,000 years. The important thing, the committee report said, is that "no fundamental, insuperable limitation to the ability of Mars to support terrestrial life has been unequivocally identified." This is really thinking big.

Some scientists extend the life zone of the sun past the four innermost, earthlike planets. Life on the giant outermost planets would have to be very different from life on earth, but a combination of theory, observation, and experimentation has produced a scenario for the creation and evolution of organisms on such planets as Jupiter.

LIFE ON JUPITER: HARD TO IMAGINE

Thinking in terrestrial terms will not get us far, because Jupiter is more like the sun than the earth. If the earth is a rock, then Jupiter is a gas. Jupiter is massive: more than 300 times the mass of the earth, more than twice as massive as all the other planets put together. The Pioneer 10 and 11 spacecraft that flew by Jupiter in 1973 and 1974 showed that the planet consists mostly of hydrogen and helium, like the sun, although methane, ammonia, and water are also present. Jupiter emits radio waves and heat, perhaps because it is undergoing the kind of gravitational collapse associated with protostars. And Jupiter spins so fast that it is visibly flattened at the poles and bulges out at the equator.

Jupiter completes one rotation every 9 hours and 55 minutes. The speed of its rotation creates the planet's distinctive bands of clouds, whose colors range from yellow to bronze. Earth's slower spin allows clouds to move north and south, producing our changeable weather. On Jupiter, the clouds are trapped into permanent bands, called zones and belts. The Great Red Spot, about whose nature astronomers argued for decades, now is known to be a more or less permanent storm raging in the upper atmosphere, kept going by the enormous forces at work on Jupiter.

When we add to this picture of giant size, gaseous composition and rapid spin two other factors, Jupiter's great distance from the sun and evidence of turbulent circulation of gases from the interior of the planet to its upper atmosphere, life on Jupiter seems difficult to imagine. But if we abandon terrestrial concepts, the unimaginable seems possible.

After all, Jupiter has what are regarded as the essentials of life: heat (at a certain level of its atmosphere), water, a reducing atmosphere of the kind that existed on the primitive earth, and organic molecules. In addition to ammonia and methane, ethane and acetylene have been found in the Jovian atmosphere, and it is possible that more complex hydrocarbons are produced there. The colors of the belts and zones could be explained by the presence of complex organic molecules; the Great Red Spot could be a collection of such compounds. Ample energy is available in the form of lightning discharges, ultraviolet light, and heat to help those compounds start on the road to life.

If there is life on Jupiter, it would have to float in the air, because Jupiter has no solid surface. However, floating organisms are not an impossibility. There is floating life of a sort on earth, in the oceans, and life in the clouds seems achievable. In 1976 a group of scientists headed by Robert L. Dimmick of the University of California at Berkeley put a terrestrial species of bacterium in a six-foot rolling drum and kept the microbes suspended in air. The bacteria were fed by blowing in mists of sugar. They managed both to eat and to reproduce while floating. In other experiments, Ponnamperuma has exposed terrestrial microbes to the great variations in temperatures and pressures that might be encountered while floating in the Jovian atmosphere. Most of them survived.

Carl Sagan and a Cornell colleague, E. E. Salpeter, recently described a possible Jovian ecology that has three kinds of hypothetical organisms living an aerial life on Jupiter: sinkers, floaters, and hunters. The sinkers reproduce as they drift downward from the upper atmosphere. They provide food for the floaters, which stay at middle levels. The hunters, which could be gaseous, low-density organisms that are miles across, move in the atmosphere by pumping out helium to achieve jet propulsion. This Jovian ecology, weird as it seems, is simply the ecology of the earth's oceans translated to the atmosphere and chemistry of Jupiter.

Saturn, while similar in composition to Jupiter, is not as good a possibility for life. Saturn is more distant from the sun, and so is colder. It does not generate as much internal heat as Jupiter. But exobiologists believe they could have a live prospect in Titan, the largest of Saturn's ten moons.

TITAN: A POSSIBILITY

Titan is big for a satellite but small for a planet; its diameter is 3,600 miles, compared to 4,200 miles for Mars. But Titan is big enough to have an atmosphere, which is known to contain methane and probably hydrogen. The atmospheric pressure on Titan is about one percent that of earth—"the most earthlike atmospheric pressure in the solar system," says Carl Sagan.

Like Mars, Titan is red. Mars is red because it is rusty, but Titan is not a rocky body like Mars. It has a low density, and it appears to consist of a mixture of water and ice. The reddish color of Titan is believed to be due to organic molecules in its atmosphere, which is similar to that of the primitive earth. And Titan is a lot warmer than it should be, apparently because of a greenhouse effect due to its atmosphere. The general belief is that the relatively high temperatures on Titan are found only in Titan's upper atmosphere, and that its surface temperatures are well below freezing. But measurements are uncertain enough to keep alive the possibility that the surface of Titan is warm by terrestrial standards.

Putting together all the could-bes and might-bes, Titan, like Jupiter, becomes an interesting spot for exobiologists. A trip to Titan is fairly high on the list of future missions that planetary scientists would like to fly. Until then, Titan is a grand subject for speculation, the stuff that science fiction scenarios are made of (like Kurt Vonnegut's *The Sirens of Titan,* which does have a habitable Titan).

So a quick sweep of the solar system for life shows one sure thing (earth), one apparent near-miss (Mars), and two outside chances (Jupiter and Titan); not a bad batting average at all. That record is something to keep in mind as we move away from the sun and consider SETI from the galactic point of view.

CHAPTER SIX

THE FORMULA

The formula was devised in 1961 by Frank Drake of Cornell. Although it really tells us mostly what we don't know, it quickly became a cornerstone of SETI, something to build on. When a pioneering international meeting on communication with extraterrestrial intelligence was held in Armenia in 1971, the participants spent a good deal of their time defining the subject by going through the formula, term by term.

It is a remarkable formula. Its purpose is to give a reasonable estimate of the number of civilizations in the galaxy which have the technological ability to communicate with other civilizations. The formula is:

$$N = R_* f_p n_e f_l f_i f_c L$$

The number of technological civilizations in the galaxy is designated as N. On the right side of the formula, R_* is the number of new stars that are formed in the galaxy every year; f_p is the fraction of stars that have planetary systems, n_e is the average number of planets in such systems that can support life; f_l is the fraction of planets on which life actually occurs; f_i is the fraction of such planets on which intelligent life arises; f_c is the fraction of those planets on which intelligent beings learn how to communicate with other civilizations; and L is the average lifetime of such technical civilizations. Multiply all the terms on the right and we have an estimate of the number of civilizations in the galaxy that are capable of communicating with other advanced civilizations.

But there is a problem in applying the formula: the estimates of the numerical values of the various terms in the equation are vague, at best. Astronomers are used to dealing with large numbers which require large margins for error; a billion or so of anything is small change in astronomy. The habit of using astronomically large and elastic numbers is common in SETI. Take the basic question of the number of stars in our galaxy. The standard answer when the question is raised is 100 billion, a nice round number. But when an argument requires it, an astronomer may casually change that to 250 billion stars. After all, it will be explained, these things are not fully understood, and either number is in the ball park.

Rough guesses are commonplace because a staggering amount of knowledge is needed for more precise measurements. At the 1971 meeting in Armenia, one of the hosts, V. A. Ambartsumian, pointed out what would be needed for a complete answer. The factor R_* is studied by astrophysics, he said, as if f_p, while n_e requires both astronomy and biology. An estimate of the value of f_l requires a knowledge of both organic chemistry and biochemistry, while f_i can be fixed precisely only with the help of neurophysiology and knowledge about the evolution of advanced organisms. As for f_c, an estimate of its value requires data from anthropology, archaeology and history. Finally, the value of L is nothing less than the major question now facing mankind. Can an advanced technological civilization of our kind survive for a long period? Will we destroy ourselves in any one of a dozen technological ways? Will we cease to be a technological civilization by turning away from technology? Not only for L, but for every factor in the equation, there are many unanswered questions. Answering them requires information from almost every scientific discipline. "It is remarkable," said Ambartsumian, "that there exists a single problem that involves so intimately subjects ranging from astrophysics and molecular biology to archaeology and politics."

FACTORS IN THE EQUATION

But even if a great deal of the needed information is not available, a start must be made somewhere. We have some data, although it is imperfect. To start with R_*, our galaxy is believed to be about 10 billion years old and contains about 100 billion stars. Therefore, an average of 10 new stars a year have formed during the lifetime of the galaxy. But that does not mean that the current rate of star formation is 10 stars a year. It can be assumed that more stars formed in the early years of the galaxy than form now, when the galaxy is mature. A reasonable estimate for the rate of formation of new stars in the galaxy today is about one a year. If this value is correct, then R_* equals one. There is plenty of margin for error in that value. An estimate of three new stars a year in our galaxy would not draw a violent argument for most astronomers. But one a year is an acceptably conservative estimate.

How many of those stars have planetary systems? Again, rough estimates are necessary. As was pointed out earlier, the evidence of stellar spin indicates that a rather large fraction of moderate-sized stars could have planetary systems. There are all those stars whose mass is about that of the sun and which spin more slowly than their masses indicate they should. In the case of the sun, astronomers are reasonably certain that the missing angular momentum was transferred from the star to its planets. But it is not at all certain that the same thing happened to all the other slow-spinning stars.

There is an alternative explanation that does not include the formation of a planetary system. Perhaps the angular momentum of a protostar is transferred to gas and dust that spirals out from the star without forming planets. Since there is almost no observational evidence of the existence of other planetary systems, we do not know how often a star lost angular momentum without gaining planets. Estimates of the fraction of sun-sized stars that have planetary systems range from one in ten to as high as five in ten. Therefore, the value of f_p could be anything from one to five.

For an estimate of the next term in the formula, the value of n_e, the fraction of planets that can support life, we have some evidence from the solar system. Earth can support life. Mars is a "maybe." Jupiter is another "maybe," although a stretch of terrestrial standards is needed. The most optimistic reading is that three of the nine planets are capable of supporting life. For the galaxy at large, we can pick an estimate starting at one planet in ten and running up to four planets in ten, as the value for n_e.

The next term, f_l, is the fraction of such planets on which life actually does arise. In setting a value for this term of the equation, exobiologists tend to be optimistic. Since the universe seems to be liberally seeded with the starting molecules of life, since living organisms appeared so early in the history of earth, and since there is no reason to believe that a planet like earth in another planetary system would undergo a different kind of early history, it is assumed that the value of f_l is high. Carl Sagan typically sets the value of n_e at one—meaning that life inevitably arises on any planet capable of supporting life. More conservative estimates are that life exists on one in every two or one in every four suitable planets.

Once primitive living organisms come into existence on a planet, what are the chances that intelligent life will evolve? There are wide differences of opinion about the value that should be set on this term of the equation, f_i. The noted paleontologist George Gaylord Simpson, an expert on evolution, has said that the chances of intelligent life arising elsewhere are very low. In effect, Simpson rejects the assumption of mediocrity. He believes that the earth is "appallingly unique," at least when we consider the long and complex sequence of events that led to the evolution of the human race.

Simpson's is a minority view. Most workers in SETI believe that while the exact sequence of events that produced the human race may be unique, the general evolutionary pathway that leads from primitive life forms to intelligent beings could be common. John Lilly, a biologist who rhapsodizes about the personality of dolphins and who did extensive work on their intelligence, believes that intelligent beings have evolved on earth by two separate pathways: one for humans, one for dolphins. (But his proposition, if true, does SETI no good because dolphins, however intelligent, are determinedly untechnological.) On the other hand, scientist-writer Isaac Asimov has said that the evolutionary pathway on another planet is bound to be so similar to that of earth that its intelligent beings will have an overall resemblance to humans—one head, two arms, two legs. Asimov reasons that the value of having a central brain, hands to do fine manipulations, and legs to maneuver with are so great that the basic design is likely to be repeated again and again.

Given this diversity of opinion, the values set by different people for f_i differ widely. The value of f_i has been set as high as one—meaning that intelligent life will evolve on any planet where life arises—and as low as .01, meaning that intelligent life will evolve on only one in 100 of such life-bearing planets. No value between one chance in a hundred and a hundred chances in a hundred need be rejected. In fact, if Lilly is correct, the value of f_i may be higher than one, since intelligent life may evolve more than once on the same planet.

Once intelligent life arises, what is the probability that a civilization will be created that has both the technological capability and the desire to communicate with other civilizations? The value that any observer sets on this factor, f_c, depends on that observer's world-view. A view that is almost standard in SETI puts technology at the peak of human achievement and adds the assumption that technology is the essential tool of an ever-curious, ever-questing, ever-expanding intelligent race. From this point of view, the rise of a technological civilization is the logical goal of any intelligent species, and the desire of such a species to move from its home planet into the frontier of the universe is unquenchable.

An alternative point of view that has gained some support in recent years holds that technology is as much a curse as a blessing. From this point of view, the 400-year-old European effort to conquer nature through science and technology is not the goal of human life but an aberration from which the human race must recover. The current estimate of the value of f_c tries to balance these two views. It is believed that only a minority of intelligent species will attempt to communicate with life on other planets. The value of f_c is set at something between one in ten and five in ten.

L=THE MAJOR QUESTION FACING MANKIND

At this point, it must be confessed that the values set for the first six terms of the right side of the equation are not very meaningful. Most of the values are so elastic that the results of multiplying them can give any result over an enormously wide range. What the people in SETI usually do is to juggle the values so that the result comes out at unity—1. This method makes the value of L the most important factor in the equation. If the value of L is high, there are many communicating civilizations in the galaxy. If the value is set low enough, we can conclude that we humans are alone—in the sense that we are the only species with both the ability and the will to communicate with other civilizations that exist at this time in the galaxy.

The last statement does not mean that the human race is the only communicating species to have existed in the history of the galaxy. In the billions of years that the universe has been in existence, millions of technological civilizations may have come into existence on planets scattered through the galaxy. But if those civilizations tend to lose their ability to communicate in a matter of centuries, the odds that two of them will be in existence at the same time is small. And since the galaxy is immense, the odds are even smaller that two such short-lived civilizations will be close enough to make contact even if they have their brief flowering at the same time. Communication with extraterrestrial intelligence is likely only if technological civilizations have lifetimes measured in many thousands of years.

As it happens, the lifetime of a typical technological civilization is not only the central issue of SETI but also the central issue of mankind today. The only technological civilization which we can study to set a value for L is ours. Mankind can end the SETI debate by destroying its technological capability, either by any one of the toxic by-products of technology or by simply losing interest in technology. In that case, the value of L is a few tens of years, and the case for SETI is closed. Or mankind could manage to fulfill the vision of a planet that maintains its prosperity and its technological drive. In that case, mankind can look forward to making contact with another advanced civilization sooner or later.

However, mankind's fate cannot be accepted as the fate of all other intelligent species that may have existed in the universe. Iosif Shklovsky, a Soviet leader in SETI, has made the point that many societies may destroy themselves in a fairly short time, but that some small fraction of intelligent societies might be able to solve the multitude of problems that face a technological civilization. These successful societies might then survive and advance for many thousands or even millions of years. At the meeting in Armenia, Shklovsky put the value of L at above 50 years and under one million years.

Shklovsky and everyone else at the meeting agreed that all estimates are no better than guesses. But an estimate of the possible number of communicating civilizations in the galaxy can be made on the bases of those guesses.

Assume that all the terms but L on the right side of the equation, when multiplied, give an answer of 1. This means that one communicating civilization comes into existence every 10 years. If the average lifetime of an advanced civilization is 10 million years, then there are about one million communicating civilizations in the galaxy. (If that number seems large, realize that it works out to one communicating civilization for every 100,000 stars.) If the average lifetime is 100,000 years—we hesitate to say "only" 100,000 years, knowing the present state of mankind—then there are 10,000 communicating civilizations in the galaxy.

Individuals in SETI have speculated almost endlessly about the value of L and the number of advanced civilizations that now exist. Those speculations, made to look more formidable by complex mathematics, can lead to almost any answer at all. Estimates have ranged from millions of communicating systems in the galaxy down to just one—us.

Just recently, these estimates have come to be of practical importance. After several decades in which the search for extraterrestrial life was financed in no formal way, scientists in the field have begun to ask for federal appropriations. When they ask for money, they must make some statements about how long the search might take. Obviously, the length of the search is directly related to the number of communicating civilizations in the galaxy. If there are millions of communicating civilizations, they are relatively close together and a relatively short period will be needed to make contact (if we know enough to go about the search in the right way). But if there are relatively few communicating civilizations in the galaxy, the effort to make contact may take almost forever, in practical terms.

"Relatively" is the important word here. "Long" and "short" have different meanings for congressmen and other laymen than for astronomers, who are used to working with very large distances and very long periods of time. A time interval that sounds reasonable to someone in SETI may impress someone outside the field as being preposterously long.

Take the estimate made by Carl Sagan at the meeting in Armenia that there might be 100,000 technical civilizations in the galaxy. As we have seen, this is a middle-of-the-road number, by the standards of the formula. To the novice, 100,000 civilizations is an encouragingly large number; the galaxy is full of them. And yet it means that the distance between civilizations is depressingly large by everyday standards; the distance would be on the order of hundreds of light-years. The time required for two civilizations to make contact would be correspondingly long. If earth were to intercept a message from a civilization at that distance tomorrow and if a return message were sent immediately, the response from the other civilization would not arrive until roughly the twenty-fifth century. But there is a more basic consideration: if civilizations are so distant from one another, the odds of them making contact are correspondingly small because the number of stars that each civilization must scan before finding intelligent life is large.

The prospects of obtaining the money for the protracted effort needed to communicate with another civilization thus would not be bright. Ronald Bracewell of Stanford University, a freewheeling member of the SETI community, makes the point that most governments are not eager to spend large amounts of money for many decades or centuries on the chance that some indefinable benefit might result in the indefinite future.

Interestingly enough, we now can begin to judge the strength of Bracewell's arguments ourselves. Requests to spend relatively modest amounts of money on a search for extraterrestrial life now are before both the American Congress and the appropriate officials in the Soviet Union. The scientists who are making the proposals believe that these first searches should be just the prelude to more long-lasting and more expensive efforts. While they are usually discreet enough not to say so directly, the scientists are fully prepared to look for decades or even centuries, and to spend billions of dollars in the search. For the time being, they are asking only for a few millions and are having limited success. If the number of communicating civilizations in the galaxy is small, the real test of Bracewell's argument will come in a few years, when the preliminary search has gotten negative results and the searchers come back to ask for more money.

Thus, if the value of N in the formula is small, the difficulty facing SETI is one of maintaining public interest and public financing for a long, long time. But if the value of N in the formula is calculated to be large, another difficulty arises—a difficulty that not only has more immediate consequences but also is more enthralling to most people.

WHERE ARE THEY?

If one assumes that there are millions of advanced civilizations in the galaxy and that the distances between them are not great—a few tens of light-years—there is an embarrassment of riches. If all those civilizations are out there, and if they have advanced technological capabilities, why haven't we heard from them by now? More than thirty years ago, physicist Enrico Fermi asked the question very bluntly: "Where are they?"

Some ingenious answers have been proposed for that question. But the simplest answer of all, and the answer that has the most attraction, can be read almost any day in a newspaper or magazine. That answer is: They're here already. (There is a variation on that theme, which has led to any number of best-selling books: They were here a while ago, but they left.)

For several decades now, observers all around the world have been sighting strange objects in the sky. Long accounts of such sightings are being published constantly. Many of those accounts include details of meetings with beings who are aboard the strange objects. A sizable portion of the American population has at least a sneaking belief that these objects may be visitors from another planet. This belief has created a large market for stories about Unidentified Flying Objects and has given birth to a subsidiary publishing industry which is based on the thesis that the reminders of ancient space travelers are all around. Since the purpose of SETI is to make contact with another advanced civilization, it seems prudent to examine the possibility that contact has already been made, either in the distant past or in in the present.

CHAPTER SEVEN

WHO GOES THERE?

On June 24, 1947, Kenneth Arnold, a businessman from Boise, Idaho, was flying his private plane over the state of Washington when he saw what appeared to be some unfamiliar aircraft near Mount Rainier. Although the aircraft were at least 20 miles away, Arnold reported later that he could see them quite clearly "flying like geese in a diagonal chainlike line, as if they were chained together." The chain was at least 5 miles long, Arnold said, and it consisted of "saucerlike things . . . flat like a piepan and so shiny they reflected the sun like a mirror."

Arnold later changed his description to say that the objects were crescent-shaped and had wings, but his early phrase, publicized by William Bequette, a reporter on the *Pendleton East Oregonian,* caught on. The "flying saucer," later to become the Unidentified Flying Object, had been born.

THEY'RE ALREADY HERE

More than three decades later, UFOs are still a topic of lively discussion on the fringe of science. The field of ufology, as its devotees call it, has several privately funded investigative groups and supports several magazines. Books on UFOs appear regularly. Ufology conventions are well attended. No day goes by without a UFO sighting, and hardly a month goes by without a long, detailed account of a human meeting with occupants of an alien spacecraft. A poll in 1976 found that some 15 million Americans have seen UFOs at one time or another and that a clear majority believes that UFOs are real. There even appears to be some underground sentiment in favor of UFO studies in the scientific community. In 1977 Peter A. Sturrock, an astrophysicist at Stanford University, found that 80 percent of the nearly 1,400 members of the American Astronomical Society who responded to a questionnaire felt that the UFO problem "certainly . . . probably . . . or possibly . . . deserves scientific study." Sixty-two of those who replied said they had witnessed an event that they thought might be related to the UFO phenomenon.

Ufology is certainly one of the more interesting phenomena of our time. Officially, the subject has been written off as unworthy of serious scientific examination. The great majority of those in SETI say that ufology belongs more in the field of mass psychology than in astronomy. (The results of Sturrock's poll can be explained as a standard example of a biased sample—that is, only those with an interest in UFOs were likely to return the questionnaire.) It is certainly true that proclaiming a belief in UFOs is not the best way to forward a career in science. The case of James MacDonald is an extreme example of that fact.

MacDonald was the senior atmospheric physicist in the University of Arizona's Department of Atmospheric Sciences. A long private study convinced him that UFOs are of extraterrestrial origin. During the 1960s, he gave a number of lectures on UFOs and even got a National Academy of Sciences grant to support his UFO research. All the while, MacDonald carried on his research in the atmospheric sciences. In 1971 he was called before a House of Representatives committee that was studying the proposed U.S. supersonic transport. MacDonald testified about his atmospheric research, which had led him to believe that the SST would damage the earth's ozone layer and thus would cause an increase in skin cancers—a finding, incidentally, which now is widely accepted.

In the questioning that followed, the discussion was steered to the UFO issue. MacDonald described his belief that there is a connection between UFO sightings and electric power failures. That statement was used to ridicule his testimony about the effect of the SST on the ozone layer. MacDonald committed suicide later that year.

WELCOME TO UFOLOGY

Ridicule remains a standard scientific response to belief in UFOs, for what scientists regard as very good reasons. One can appreciate those reasons by stepping into the world of ufology. From the outside, ufology may appear to be a sober subject. On the inside, it is a fabulous garden of exotic growths, full of a lush underbrush with many hidden traps. Ufology is not a unified movement, to say the least. It has its conservative wing, which is dedicated to proving the extraterrestrial hypothesis, and any number of other wings that can be difficult to describe. In a UFO magazine, technical reports on the authenticity of UFO pictures rub shoulders with articles on "How to Be Captured by UFOlk" (". . . make yourself available. Many UFOlk can read your mind. They know if you're willing.") and reports of miraculous healings effected by humanoids. Proposals that UFOs are manifestations of demonic activity get the same grave consideration as theories about the planetary origin of extraterrestrial visitors. Books proposing that the earth is hollow and that UFOs come from the interior are reviewed. Advertisements invite readers to "unlock your hidden powers" or to send away for "over 100 ready-to-use mystic chants for money, power and love."

In this atmosphere, it is difficult to recall the early Cold War days when UFOs were taken seriously as a possible threat to U.S. national security. In the rash of sightings that followed Kenneth Arnold's original report, the Air Force set up what it called Project Saucer to investigate UFOs. It was a time when Americans had fresh memories of a war in which secret weapons had been quite real: the V-2 rocket, the jet plane, the atomic bomb. For a time, a security classification was placed on the military investigation of UFOs.

THE FIRST "PROJECTS": SAUCER, SIGN, GRUDGE, BLUE BOOK

The urgency wore off as the years passed. Project Saucer gave way to Project Sign, which was followed by Project Grudge and then by Project Blue Book. For a time, at least some Air Force investigators believed that the extraterrestrial hypothesis was tenable. That time passed, as familiarity with the objects grew. By the early 1960s, the Air Force was being criticized about UFOs from both sides: some felt that the Air Force was not devoting enough effort to the study, others that UFOs were not worth an effort at all.

In 1966, the Air Force decided to let an independent scientific investigation settle the issue. The man chosen to lead the inquiry was Edward U. Condon of the University of Colorado, one of the most eminent physicists in the country. Condon clearly was doing himself no good by agreeing to take on the job. As a 1966 letter from Thurston E. Manning, a vice-president of the university, explained, "there was an understandable reluctance on the part of many scientists to undertake such a study . . . the subject had achieved considerable notoriety over the years."

Condon, who had already proved his toughness by surviving a smear by the House Committee on Un-American Activities during the McCarthy years, got all the trouble that could have been anticipated. Part of the trouble was due to the university's obvious distaste for the project. Robert Low, assistant dean of the university's graduate school, wrote a memo about the possible effect the Condon study might have on the university's image. The memo read in part:

> In order to undertake such a project, one has to approach it objectively. That is, one has to admit the possibility that such things exist. It is not respectable to give serious consideration to such a possibility. . . . We would lose more in prestige in the scientific community than we could possibly gain by undertaking the investigation. . . . Our study would be conducted almost exclusively by nonbelievers who, although they couldn't

possibly prove a negative result, could and probably would add an impressive body of evidence that there is no reality to the observations. The trick would be, I think, to describe the project so that, to the public, it would appear a totally objective study, but, to the scientific community, would represent the image of a group of nonbelievers trying their best to be objective but having an almost zero expectation of finding a saucer . . .

The memo was found in the project's files by Norman E. Levine, a member of the staff. He showed it to David R. Saunders, one of the psychologists on the staff. Saunders was a member of NICAP, the National Investigative Committee on Aerial Phenomena, a private organization devoted to proving the extraterrestrial hypothesis. Levine and Saunders passed the memo on to NICAP, and it soon was published. Condon fired both Levine and Saunders for leaking the document.

THE CONDON REPORT

Criticism from UFO believers was inevitable. Condon's final report, issued as the *Scientific Study of Unidentified Flying Objects* on January 8, 1969, increased the criticism, in the second paragraph of the report, which ran nearly a thousand pages, Condon said:

> Our general conclusion is that nothing that has come from the study of UFOs in the past 21 years has added to scientific knowledge. Careful consideration of the record as it is available to us leads us to conclude that further extensive study of UFOs probably cannot be justified in the expectation that science will be advanced thereby.

Before the report was issued, the Air Force had it evaluated by a committee of the National Academy of Sciences. The special committee said in its report:

> We are unanimous in the opinion that this has been a very creditable effort to apply objectively the relevant techniques of

> science to the solution of the UFO problem. . . . On the basis of present knowledge the least likely explanation of UFOs is the hypothesis of extraterrestrial visitations by intelligent beings.

The Air Force then announced that it would close down Project Blue Book and get out of the UFO investigation business. Officially, that ended government action on UFOs. For example, if you call the National Aeronautics and Space Administration with a UFO report today, you will be referred politely to NICAP. Unofficially, the world of ufology teems with rumors that one government agency or another—the FBI, the CIA, or someone—is still active in the UFO field. One recurring theme in ufological discussions is that of the Man in Black, an ominously threatening caller from an unidentified government agency who uses threats to silence observers of particularly exciting UFO events. Few ufologists can name someone who has actually spoken to a Man in Black; generally, the menacing figure is glimpsed ducking around a corner or dodging behind a pole, at some distance. Nevertheless, the Man in Black is believed to be a constant presence.

Although Condon took pains to point out that he had sought the widest possible scientific support, every aspect of his report was criticized by ufologists. The flavor of the criticism can be described by the title of the *Look* magazine article that John G. Fuller, a prolific writer, based on the leaked Low memorandum: "Flying Saucer Fiasco; The Extraordinary Story of the Half-Million-Dollar 'Trick' to Make Americans Believe the Condon Committee Was Conducting an Objective Investigation."

The heart of the Condon report was a case-by-case study of UFO sightings. Ufologists have two criticisms of the studies: first, that Condon chose only cases that could be easily explained; and second, that many of the cases are listed as being unexplained. Most ufologists say they are willing to concede that 90 percent or more of UFO sightings can be explained by natural causes. But they dispute the Condon thesis that the other 10 percent or so could also be explained by natural causes, given enough time and effort. Ufologists rest their case on (a) the small percentage of UFO sightings which they say have no possible natural explanation, and (b) a body of what are described as reliable reports of personal contacts with extraterrestrial creatures.

Opinions differ about the personal-contact cases. NICAP tends to be skeptical about them. But APRO, the Aerial Phenomena Research Organization, headed by James and Coral Lorenzen of Tucson, is less conservative. "Abduction cases are at least mind-boggling," James Lorenzen told one interviewer. Of one personal contact report, he said, "I don't know if it came from another civilization, but a human was aboard something under very unusual circumstances."

One's first reaction is to be impressed by the number of UFO sightings that ufologists say are unexplainable by usual means. A computer bank at the Center for UFO Studies, established by J. Allen Hynek in Chicago, lists 80,000 such cases, including 800 personal-contact stories.

Hynek's own story, which is entirely terrestrial, is a history of the UFO movement in itself. He appeared on the UFO scene in the very earliest days, when he was an astrophysicist on the faculty of Ohio State University; the Air Force hired him to investigate the Arnold report and the early sightings that followed. He remained an Air Force consultant on UFOs until the Condon report ended the Air Force investigation of UFOs. Hynek then became what can best be described as a UFO free-lancer, founding the center, writing and talking on the subject frequently. Most recently, he became prominent as technical adviser to the film, *Close Encounters of the Third Kind.*

In the 1950s and the 1960s, Hynek stressed the possible extraterrestrial origin of UFOs. In the 1970s, he has widened the possibilities. A book which Hynek wrote in 1972, *The UFO Experience: A Scientific Inquiry,* and which proposed a major international effort to gather and study data about UFOs, contained a mention of "extraterrestrial visitors or the more esoteric notions of time travel or parallel universes" as equally eligible explanations of UFOs. A later book, *The Edge of Reality,* which Hynek wrote with another long-time ufologist, Jacques Vallee, leaned more toward a psychic explanation. In an interview, Hynek spoke of UFOs as a possible link between "two parallel realities."

WILDER MUSIC: THE UFO CRAZE

And that is one of the problems in trying to evaluate the extraterrestrial hypothesis of UFOs: wilder music is being played in the background. At the First International UFO Congress, held in Chicago in June 1977, most of the sessions were devoted to the extraterrestrial hypothesis. But there was a "symposium on contactee phenomena" that included talks titled "The Contactees' Psychic World," "The Virgin Mary—Ufonaut Extraordinary," "UFOs and the Fairy Faith," and "Revelations of the Space Brothers." Jacques Vallee, Hynek's collaborator in writing, delivered the talk on the "fairy faith" as well as a lecture outlining his belief that "human beings are controlled by some sort of superhuman consciousness somehow involved with the UFO problem."

Hynek spoke twice at the congress, including a summary address on "The Future of Ufology." Kenneth Arnold, billed as "the man who started it all" was there. Another session was on George Adamski, who maintained that he met earthlike visitors from Venus in the California desert in 1952, that in a later meeting he was taken aboard a spacecraft by two other space visitors, also earthlike, who identified themselves as being from Mars and Saturn, respectively, and that they all flew to a "mother ship" that took him to the moon, "in which vegetation, trees, and animals thrive, and in which people live in comfort."

But this sort of bizarre buncombe is not the face that ufology usually presents to outsiders. On the rather rare occasions when UFOs break into scientific forums—usually when a believer's letter is published in a journal—the psychic aspects of ufology generally are slighted. Investigations by nonbelievers also ignore such possibilities as parallel realities and psychic phenomena, concentrating on the extraterrestrial hypothesis.

Investigations of UFO reports by people outside the world of ufology are not common. Most scientists ignore the subject, on the grounds that they have much better ways to spend their time. (One notable exception was the late Donald H. Menzel, a distinguished astronomer at the Harvard College Observatory who spent a great deal of time on UFOs and whose last book, *The UFO Enigma*, written with Ernest H. Taves, summed up his reasons for skepticism.) Only a few nonbelievers are willing to spend the time and energy needed to check out UFO reports. Perhaps the leading skeptical investigator of UFO cases is Philip J. Klass, senior avionics editor for *Aviation Week and Space Technology* magazine. A combination of technical knowledge and diligence makes Klass unusually qualified for his self-chosen work.

CASE HISTORIES

Two cases will give an idea of how exhaustively Klass can go into a report of a UFO sighting. One is the case of the UFO that reportedly was seen near McMinnville, Oregon, on May 11, 1950, by a Mr. and Mrs. Paul Trent. The Trents said they had taken two pictures of the objects. William K. Hartmann, the chief photoanalyst for the Condon study, examined the pictures and reported: "This is one of the few UFO reports in which all factors investigated, geometric, psychological and physical, appear to be consistent with the assertion that an extraordinary flying object, silvery, metallic, disc-shaped, tens of meters in diameter, and evidently artificial, flew within sight of two witnesses."

Nearly twenty years after the McMinnville incident, Klass obtained copies of the two photos. They were examined by Robert Schaeffer, a colleague of Klass's, who established: (a) that the shadows in the pictures indicated that they had been taken in the early morning, rather than in the late afternoon as the Trents said, and that (b) one picture which the Trents said had been taken a few seconds before the other had actually been taken several minutes earlier. Klass took the analysis to Hartmann, who said that the new information "removes the McMinnville case from consideration as evidence for the existence of disclike artificial aircraft," his phrase for UFOs.

An even more unusual case was the incident in which an Air Force RB-47 jet bomber apparently was followed for 90 minutes and 700 miles as it flew from Mississippi to Oklahoma on July 17,

1957. Not only were the crew members able to see the UFO as a bright light in the early-morning sky, but the object also was detected on radar. A long and exhaustive investigation by Klass has led him to believe that the radar signals picked up by the RB-47's equipment actually originated with ground-based radar stations in the area, and that the lights seen at various times by the bomber's crew were a meteorite, the star Vega, and the lights of a commercial airliner landing at the Dallas Airport.

UFO believers have two reactions to Klass's debunking efforts: either they ignore them (the McMinnville case and the RB-47 incident show up regularly in new UFO books as unexplained sightings) or they substitute other sighting reports. There are always plenty of other reports; Klass himself acknowledges that in the time that it takes him to investigate one case, a dozen or more will be discovered by ufologists. His response to the ufologists is to put his money where his mouth is. Klass has offered to buy back every copy of his book, *UFOs Explained,* at its list price of $8.95 "if at any time evidence of an authentic extraterrestrial spaceship is ever found." He also has a standing offer: if anyone agrees to pay him $100 a year for ten years, Klass promises to pay $10,000 if UFOs are proven to come from an extraterrestrial civilization. As of 1977, he had two takers for the $10,000 bet. Most ufologists seem to agree with James Lorenzen of APRO, who told one interviewer that Klass's offer is "a tricky and devious proposition which typifies his whole approach to the UFO problem."

THE CHARIOTS OF VON DÄNIKEN

Klass has his hands so full with current UFO reports that he has never taken on one of the most successful offshoots of the subject: Erich von Däniken and his prehistoric visitors from extraterrestrial civilizations. Von Däniken has been remarkably successful, at least with general readers. His first book, *Chariots of the Gods?* published in 1969, sold millions of copies in Europe and here. His next book, *Gods from Outer Space,* and its successors continued the best-selling record. Indeed, von Däniken seems to have started a publishing industry. Any paperback store features a shelfful of books dedicated to the proposition that extraterrestrial visitors were here thousands of years ago and that they played an important role in (depending on the tack of the author) either starting life on earth or establishing civilization here.

Von Däniken's success puzzles many scientists who maintain that he is neither original nor very imaginative. In the early decades of the century, an eccentric American writer named Charles Fort proposed that extraterrestrial beings had visited the earth—indeed, that earthlings were unwitting property of powerful, unseen beings. Fort's books never had a fraction of the success that von Däniken has achieved, but Fort was there first with the idea.

In addition, scientists who have read von Däniken's books say that there is hardly a correct idea in any of them. They accuse von Däniken of naïveté, at the very least. It has been pointed out that one of von Däniken's basic methods has been the systematic underestimation of the technical ability of early civilizations. For example, von Däniken says that the natives of Easter Island could never have carved the huge statues that dot the island. He supports that belief with a report of his own failure in trying to carve the island's volcanic rock, and concludes that only "alien visitors" could have supplied the islanders with the "sophisticated technical tools" needed to make the statues.

However, anthropologists familiar with the Pacific say that many Polynesian peoples carve large funerary statues. On most islands, the carvings are of wood. Easter Island is almost treeless, but it has an ample supply of volcanic rock. Instead of being evidence of alien visitations, the Easter Island statues thus are just one example of a common local activity.

A few critics have tracked down von Däniken's interpretations, one by one. Where he marvels at engravings of rocketlike machines in early China, others point out that the Chinese shot off rockets in festive celebrations. Where he expresses amazement that drawings from primitive South America show camels and lions that did not exist on the continent, others point out that llamas and pumas did. Looking at a carving in the Mayan site of Palenque, in what is now Mexico, von Däniken sees a man dressed in tight pants and socks, sitting at the controls of a spaceship. Anthropologists see a man wearing a conventional Mayan loincloth and headdress and lying on a sacrificial altar—a common theme in Mayan art.

Often, very little detective work is needed to refute one of von Däniken's ideas. In *Chariots of the Gods?*, for example, he heaps scorn on the thought that the ancient Egyptians could have built the pyramids some 4,500 years ago without extraterrestrial help. They simply could not have moved the massive blocks of stone of which the pyramids are made, von Däniken said, unless visitors from space gave them assistance. He scoffs at the belief that the stone blocks were moved on wooden rollers and were pulled by ropes. Von Däniken says bluntly that the Egyptians did not have trees from which to make rollers and that they did not know of rope.

Egyptologists don't know where von Däniken gets his information. They point out that there are ample records of a timber trade which brought cedars from Lebanon in large quantity. As for rope, archaeologists have more than pictures of rope being used by the ancient Egyptians; they have some of the actual rope, including samples found in the quarries from which the stones of the pyramids were taken.

Similar examples of farfetched conclusions from the writings of von Däniken and others of his school are almost endless. Nevertheless, their books sell marvelously well, while the demand for skeptical writings is minimal. Those scientists who have taken the time to study the subject often explain the success of such books as a response to the gap that the decline of religious beliefs has left in our lives.

Religious belief was once a unifying factor in society. Science, which has done so much to undermine religion, and which has made life more difficult by substituting constant change for old-time stability, offers no comforting substitutes for the old verities. Pseudo-science does. The "little green men" who step out of UFOs usually have messages of comfort for mankind. The ancient visitors of von Däniken are powerful beings who descend from the sky to offer help to a bewildered human race. Once again, Someone Up There is watching over us. UFOs, present or past, offer a scientific basis for faith and hope.

THE RESPONSE OF ASTRONOMY

Unfortunately, the hard numbers of astronomy do not provide much support for believing that UFOs are extraterrestrial visitors. There are three problems in accepting the extraterrestrial hypothesis of UFOs: the galaxy is very large, it is very old, and the cost of interstellar travel is very high.

Assume that there are a million civilizations in the galaxy, and that all of them are capable of interstellar travel. Assume that all of these civilizations are both willing and able to explore the galaxy systematically. There are about 100 billion stars in the galaxy, and about 10 billion of them are candidates for planetary systems that could support life. The sun is one of the 10 billion. Since the sun does not seem to be unusual in any way, it probably would not be high on the list of priorities of an exploring civilization; it would have to wait its turn in a methodical search.

If those one million civilizations each launch one starship a year, and if each starship visits one planetary system in a year—both optimistic assumptions—it will take a total of 10,000 years before one starship visits our planetary system. To put it another way, the earth could expect one visit from a starship every 10,000 years. Or, to put it yet another way, if the earth is visited by only one starship a year, each of the one million exploring civilizations must be launching 10,000 starships a year!

But ufologists are not talking about just one starship visit a year. If even a fraction of the reports that they have collected are true, the earth is being visited constantly by what amounts to a swarm of extraterrestrial spaceships. The sheer number of UFO reports is one reason why people in the SETI community do not believe in them. One is reminded of what Albert Einstein is reported to have said when told that a hundred scientists had signed a statement saying that relativity was wrong. Einstein said: "If I were wrong, one would be enough."

One sound, indisputable case would be enough to establish the extraterrestrial origin of UFOs. In the opinion of just about everyone in the SETI community, there is no such case.

Carl Sagan, a leading spokesman for SETI, always says the same thing on those frequent occasions when he is asked about UFOs: there are no cases that are simultaneously reliable—meaning that the object has been seen by a large number of good witnesses—and exotic, meaning that it cannot be explained by ordinary means. And Sagan can certainly not be accused either of narrowness or of lack of imagination. Indeed, his willingness to entertain imaginative ideas about extraterrestrial life is well known.

Another reason why people in the SETI community are nonbelievers in UFOs is their knowledge about the nature of interstellar travel. As we shall see shortly, traveling between stars is an expensive, long-range effort. If an alien spaceship arrived on earth, it would have come a very long distance. It will also be making a rare discovery; on the average, it will have visited 9,999 other stars without success, if our earlier calculations are correct. Extraterrestrial visitors could thus be expected to make every effort to establish contact quickly with the scientific and political leaders of earth. But if the ufologists are right, the starships that have come so far to make contact have been flying around earth rather aimlessly for several decades, making themselves known to everyone but a qualified scientist or political leader. Loggers, housewives, blue-collar workmen—all have been contacted (if the UFO stories are true) and have even been aboard extraterrestrial ships, but nothing has been done to carry out what must be the main mission of any interstellar visit.

CLOSE ENCOUNTERS OF A QUESTIONABLE KIND

In short, UFOs simply do not make sense to the people in the SETI community. "To be quite truthful," said John Billingham of the Ames Research Center, "we do not pay significant attention to the UFO issue at all. We feel that the whole area is so debatable, uncertain, and unscientific that it is probably not going to help anybody very much. We don't do anything on UFOs, and we do not recommend that anybody else do anything on them."

Nevertheless, public interest does keep the issue alive. The best example of that phenomenon is the success of *Close Encounters of the Third Kind,* a science fiction film whose concluding scene is a meeting between earth's scientists and a huge spaceship carrying (literally) little green men.

The movie was released, with the usual publicity building, late in 1977. A few months earlier, Philip Klass was predicting that there would be a new wave of UFO sightings that would coincide with the release of the film. That wave—a "UFO flap"—came along right on schedule. Less predictably, the rising interest in UFOs brought a request from the White House for a new NASA investigation of UFOs. (Before he became president, Jimmy Carter saw a UFO. Scientists who have studied his report are satisfied that he saw the planet Venus.) In addition, there was a report in the magazine *U.S. News and World Report* early in 1977 saying that the government was about to open its secret files to publish some previously restricted information on UFOs. Finally, Sir Eric Gairy, the prime minister of Granada, made one of his periodic appeals to the United Nations on October 7, 1977, for creation of a UN body to study the UFO question.

Nothing came of all of this. As for secret information in government files, Frank Press, President Carter's science adviser, was told by both the Central Intelligence Agency and the Air Force that there wasn't any. In the UN, Gairy spoke to a bored General Assembly, most of whose members had walked out by the time he finished. No one bothered to respond, and a draft resolution on UFOs drew almost no reaction. The White House request to NASA prompted a press release in which the space agency said that its scientists would "keep an open mind and maintain our scientific curiosity and a willingness to analyze technical problems within our competence," but said that:

> There is an absence of tangible evidence available for thorough laboratory analysis and because of the absence of such evidence we have not been able to devise a sound scientific procedure to investigate the phenomenon. To proceed on a research task without a disciplinary framework or an exploratory technique in mind to find ways in which this problem can be studied would be wasteful and unproductive. . . . I would therefore propose that NASA take no steps to establish a research activity in this area. . . .

Despite such blunt words, the UFO phenomenon rolls along briskly outside the scientific community. Hynek's books enjoyed a surge in sales, and Hynek himself was a frequent visitor to television talk shows. More people kept seeing flying objects. An organization called Ground Saucer Watch, Inc., sued the CIA for release of all the agency's secret records on UFOs. And the success of *Close Encounters of the Third Kind* and another science-fiction film, *Star Wars,* started a wave of science fiction in Hollywood. The feeling that something important was just around the corner remained alive in ufology, as it has been for decades. And as it probably will be for many more decades.

CHAPTER EIGHT

TRAVELING

People in SETI can be divided into two groups: listeners and travelers. Listeners believe that interstellar travel is so difficult and costly that the only practical method of contacting an extraterrestrial civilization is by using radio messages. Travelers believe that voyages between stars are practical for advanced civilizations. Travelers are in the minority.

The itinerary of Pioneer 10, a spacecraft that is now traveling through the solar system, helps illustrate the reason for doubt about the ability of earth to make physical contact with beings on a planet orbiting another star. Pioneer 10 was launched from earth on a mission to explore the region of space near Jupiter in March 1972. It arrived at Jupiter in December 1973. Traveling at a speed of 35,000 miles per hour, it will not leave the solar system until the 1980s. (Pioneer 11, launched a year later than its sister ship, was targeted for a 1979 rendezvous with Saturn after its 1974 encounter with Jupiter.) Pioneer 10 carries a plaque designed to inform inhabitants of other planets that intelligent life exists on earth, and to enable such beings to locate earth. At its present rate of speed, Pioneer 10 will arrive at the distance of the nearest star at 80,000 A.D. Its projected interstellar travel time is thus many millennia longer than civilization has existed on earth. Those who sent the message aboard Pioneer 10 have no expectation at all of being alive when and if an answer is received.

INTERSTELLAR VOYAGES: A QUESTION OF PROPULSION

The two Pioneer spacecraft were launched with chemical rockets, the only kind that the human race has thus far used to launch spacecraft. Less patience on the part of senders of an interstellar message is needed if more powerful methods of propulsion are used. One such method is the use of thermonuclear bombs, which got fairly serious consideration in the 1960s under the name of Project Orion. The principle is simple: bombs would be expelled from the rear of a spacecraft and would explode. The force of the explosions would accelerate the spacecraft. Repeated explosions would propel the spacecraft at a reasonable fraction of the speed of light.

Freeman J. Dyson, a theoretical physicist at the Institute for Advanced Study in Princeton, New Jersey, and the source of many provocative ideas in SETI, has calculated that an interstellar voyage could be made by a spacecraft that started out with 30,000 hydrogen bombs. If one bomb were exploded every second for three days, Dyson calculates, the spacecraft would achieve a speed of about 3 percent the speed of light. The shock to the crew of the spacecraft would be absorbed by a giant plate mounted at the rear of the craft. When all the bombs were exploded, the spacecraft would have lost six-sevenths of its original mass. It would arrive at Alpha Centauri, a binary star system that is the sun's nearest stellar neighbor, 4.3 light-years distant, in about 130 years.

Several problems present themselves. When the spacecraft arrived at Alpha Centauri, it would be traveling at a speed of more than 5,500 miles per second, which would not be ideal for detailed observation of the star system. To slow the spacecraft or to turn it back toward earth would require another large investment in hydrogen bombs. These bombs would have to be carried all the way from earth and would represent a large fraction of the spacecraft's payload. The trip would not be cheap.

SPACE ARK: AN ECOLOGICAL COMMUNITY

Members of the crew would have to be a special kind of people. The most likely possibility they face is a lifetime aboard the interstellar spacecraft, with the hope that their children or grandchildren would see Alpha Centauri up close. A spacecraft that set out on a journey of more than a century would have to be of an unusual design. It could not meet the needs of its passengers by ordinary stores. Instead, it would have to be large enough to contain a complete ecological community whose flora and fauna could sustain a life-to-death cycle for generations. As early as 1929, J. D. Bernal envisioned such a "space ark" setting off on an interstellar voyage. Those who boarded the ark as it cast off from earth would be aware that they would never see their native planet again. They would also be aware that they were entrusting their lives to a machine that would have to maintain near-perfect functioning for many decades, far longer than any machine of comparable complexity has operated. The psychology of such an interstellar exploratory trip is interesting in the extreme.

Hydrogen bombs are not necessary for an interstellar trip, according to some studies. The British Planetary Society has a concept called Project Daedalus, in which an interstellar probe with a 500-ton payload would travel to Barnard's star. The starting weight of the probe would be 68,620 tons, most of which would be fuel consisting of small pellets of deuterium (an isotope of hydrogen whose nucleus consists of one proton and one neutron) and helium-3 (two protons, one neutron). The pellets would be fired into a strong magnetic field at the rate of 250 times a second. There they would be heated and compressed by electron beams that would cause them to fuse, releasing enormous amounts of energy. A 400-foot shell would absorb the energy to propel the spacecraft at a velocity that would eventually be 14 percent the speed of light.

The 5.9-light-year journey to Barnard's star would take 47 years in such a spacecraft. The spacecraft would fly by Barnard's star in less than 10 hours, releasing ten or twenty probes as it passed. Messages from those probes, describing the environment near Barnard's star, would be received on earth nearly six years later. The mother ship would hurtle into the galaxy, lost to earth forever.

ENZMANN'S "FLYING ICEBERG"

A comparable interstellar spacecraft project that leaves open the possibility of a return to earth was described in 1964 by Robert D. Enzmann of the Raytheon Corporation. The Enzmann spacecraft would be a huge, strange device: a cylinder 300 feet in diameter and 1,000 feet long, with a ball of frozen deuterium 1,000 feet in diameter at one end. The deuterium would not only serve as fuel for the fusion engines, but would also provide protection against radiation from the engine. The spacecraft would set off with a complement of 200 and would have room for 2,000, assuming a steady growth in population during the trip. A return voyage would be possible if the spacecraft could replenish its supply of deuterium on arrival in a new planetary system. Enzmann's "flying iceberg" would be able to accelerate to 10 percent of the speed of light and would reach Alpha Centauri in slightly more than 50 years.

These numbers are not regarded with happiness by the "listeners" in SETI. A fifty-year one-way flight would require special measures, many of which are only theoretically possible at the current state of scientific knowledge. The simplest concept—the "space ark"—would require the creation of a fully balanced ecological community that could survive without outside help in the hostile environment of space for decades. No such closed system has been produced by mankind.

Another proposal provides a standard spacecraft environment, but suggests genetic engineering or similarly advanced biological techniques to slow down the metabolic requirements of the crew, reducing their demand for food, water, and air. The science-fiction movie *2001, A Space Odyssey,* had such a plot device. Even though the spacecraft of the film had a mission that kept it within the solar system, most of its crew members were put into hibernation—suspended animation—for the trip. In the film, they were done in by the deranged master computer, HAL, which was deactivated by the survivors. Suspended animation for interstellar travel is still a movie concept.

Rather than fooling around with halfway measures such as suspended animation and fusion-powered interstellar probes, some SETI theorists prefer to demonstrate the difficulty of interstellar travel by leaping to the outermost limits of current concepts. As far as is known, the best rocket that physical law will allow is the photon rocket, which converts matter into energy with 100 percent efficiency. (By contrast, fusion power converts only 0.4 percent of matter into energy.)

The fuel for a photon rocket would consist of equal amounts of matter and antimatter, which annihilate each other on contact and flash into photons. It has been calculated that a photon rocketship could make a round trip from earth to Alpha Centauri in about ten years. The takeoff weight of the spacecraft would be 34,000 tons, of which 33,000 tons would be fuel. A recent NASA study estimated the cost of the fuel for one voyage at $100 million billion and added: "To discover life we might have to make many thousands of such sorties."

Cost is only one problem of the photon engine. A large quantity of heat would be released by the matter-antimatter reaction. The NASA study set the initial power of the engine at a billion billion watts. If only a millionth part of this energy were absorbed by the spacecraft, the cooling surface needed to dissipate the excess heat would be about 1,000 square miles. The study also noted "the problem of interstellar dust, each grain of which becomes a miniature atomic bomb when intercepted at nearly optic velocity." (Optic velocity is the speed of light.)

OF LASERS, "RAMJETS," AND "HYPERSPACE"

There are other proposals for interstellar travel. One is to use powerful earth-based lasers to push spacecraft along. Lasers emit light of an unusual kind. Light consists of waves. The light from an ordinary lamp or from the sun consists of a mixture of waves of different lengths. A laser emits light of one wavelength. Laser light is thus much more powerfully focused than ordinary light. The theory is that a laser could be focused on a spacecraft, which would be propelled by the energy in the light beam. There is nothing wrong with the theory. However, such a spacecraft would require a laser that is much more powerful than anything yet conceived—let alone built—on earth. If planetary exploration is the objective, lasers of equal power would have to be available at the other end of the journey to slow down the spacecraft. A laser-propelled spacecraft might be a possibility in the distant future; today it is only a concept.

Another concept that has attracted a good deal of attention is the interstellar ramjet, which was proposed in 1960 by engineer Robert W. Bussard. A ramjet is a jet engine that compresses air by the force of its forward motion, then expels the air to produce thrust. An interstellar ramjet would be fueled by gathering in the thin mist of hydrogen that is found in outer space, between stars. As Bussard envisioned it, the interstellar ramjet would first be accelerated to a high speed by conventional means, such as ordinary chemical rockets of the kind used today. At that time, the ramjet would take over. It would scoop in the interstellar gas and compress it to the point where the hydrogen atoms would fuse. This fusion energy would produce a high-velocity exhaust that would drive the spacecraft. Bussard calculated that an interstellar ramjet could achieve a steady acceleration of 1 g—that is, about the gravitational pull of earth. Such an acceleration would enable the spacecraft to reach a velocity near the speed of light in a matter of months.

The speed of light is a desirable objective for the design of an interstellar spacecraft. A ship traveling nearly the speed of light could travel to nearby stars in time periods now proposed for manned flights within the solar system. Alpha Centauri would be about four years away, Barnard's star about six years away (double those times for round trips). Neither a space ark nor suspended animation would be needed for such journeys.

However, there is a difficulty with Bussard's interstellar ramjet. It is easy to propose but difficult to build. Hydrogen is not plentiful, by terrestrial standards, in outer space, so a large scoop would have to be built to get enough of it to fuel the spacecraft. Bussard calculated that in regions where interstellar hydrogen is relatively dense by outer-space standards, a scoop more than 60 miles in diameter would be needed for a 1,000-ton ship. The combination of light weight and large area for the scoop creates problems. A scoop of that diameter built of mylar plastic one mil thick would weigh 250,000 tons. The need to build such a scoop could be avoided by using a magnetic field to gather the hydrogen, but there is no clear idea of how to produce a field of that size. Bussard's original proposal acknowledged that the fusion reactor needed for the ramjet spacecraft is beyond present capabilities. So is the spacecraft's propulsion system. But he had a hopeful note: "There is likewise no reason to assume that such a device is forever impossible."

But the listeners in SETI say that such a spacecraft would pose serious problems even if the scoop, the fusion reactor, and the propulsion system could be built. For example, how is the spacecraft to be decelerated when it arrives at its target star? The ramjet has no brakes, and turning it around would not be easy. And as the spacecraft accelerates toward the speed of light, the stars toward which it is heading will start to disappear as the motion of the ship shifts their light toward shorter wavelengths. As it gets close to the speed of light, the spacecraft will have to be piloted by a crew that is looking into a growing circle of blackness dead ahead. Finally, to a spacecraft traveling at great speeds, grains of space dust and interstellar gases are high-speed missiles capable of doing damage.

As outlandish as the interstellar ramjet may seem, there are even more speculative proposals for interstellar travel. One of them is to first discover a passage through "hyperspace," a theoretical passageway through another dimension that would enable a spacecraft to overcome the speed-of-light barrier. Another scheme would convert both spacecraft and crew into some form of radiant energy, transmit the energy to distant stars and then reassemble the energy into matter, more or less in the way that telephone messages now are transmitted, or the way the "Star Trek" crew is beamed up and down from the *Enterprise*. Yet another theory is that interstellar travel might be possible using tachyons: postulated particles that travel faster than light. However, the existence of tachyons is still highly speculative.

IS INTERSTELLAR FLIGHT "OUT OF THE QUESTION"?

Looking over the list of proposals for interstellar travel helps one to understand why the listeners are dominant in SETI today. At least two NASA reports on SETI, prepared by committees on which listeners predominated, contained the same conclusion, word for word, on interstellar travel:

> A sober appraisal of all the methods so far proposed forces one to the conclusion that interstellar flight is out of the question not only for the present but also for an indefinitely long time in the future. It is not a physical impossibility but it is an economic impossibility at the present time. Some unforeseeable breakthroughs must occur before man can physically travel to the stars.

Nevertheless, pockets of traveler sentiment exist within the SETI community. One traveler is Ronald Bracewell of Stanford University. Just as listeners believe that interstellar travel is impractical now, Bracewell believes that listening for signals from an extraterrestrial civilization is impractical for three reasons: a listening program will cost too much, it will take too much time to be acceptable to political leaders, and it has an unacceptably low chance of succeeding unless the universe is rather thickly packed with advanced civilizations. Bracewell has said that contact with another civilization will probably be established not by sending radio waves, nor by sending manned spacecraft, but rather by the use of unmanned interstellar probes.

SENDING UNMANNED PROBES TO A "LIKELY STAR"

Just as we have sent unmanned probes to Jupiter and to Mars, Bracewell says, the time will come when we send unmanned probes to the stars. As is common in SETI, Bracewell assumed uninterrupted technological growth. He anticipates that terrestrial scientists will someday have the ability not only to send spacecraft to the vicinity of a "likely star," but also to put the probe into orbit in the star's habitable zone and to have it seek out a likely planet in the life zone. If the planet has no advanced life, the probe can fulfill a scientific function by making the same sort of measurements as Viking made of Mars. If there is a technological civilization on the planet, the probe can establish contact and begin a dialogue between civilizations.

How could the probe make contact? Bracewell favors a method that is brilliantly simple. It requires no more than an ordinary radio or television receiver hooked up to an inexpensive transmitter. When the probe detected a radio television message from the planet, it would wait a few seconds and then start transmitting the same message. For example, if the probe happened to tune in on a classical music station that was broadcasting Beethoven's Fifth Symphony, it would transmit the music with a time lag of ten or fifteen seconds. The time lag would be selected to distinguish it from all known natural causes of repeat signals, such as radio echoes from the ionosphere, the band of charged particles in the earth's upper atmosphere.

If the first transmission did not elicit a response, the probe would try the same tactic on another frequency. Sooner or later, Bracewell believes, engineers and scientists on the planet are bound to recognize the nature of the signal and to work out the orbit of the probe. At that point, the interstellar messenger would be programmed to (a) inform its home planet that contact with an advanced civilization had been made and (b) open a cautious dialogue leading to a full-fledged exchange of information that would allow the two home planets to make direct contact. (The opening phase of the dialogue would have to be cautious to avoid blundering intervention into a delicate international situation of the sort that now exists on earth.) For example, the probe could carry a computer whose memory would contain television or radio messages that would give its home planet's position, describe the planet's inhabitants and their language, and give instructions on the best way to make direct planet-to-planet contact.

If the universe has a fairly large population of technical civilizations, Bracewell says, such interstellar probes would not be necessary because advanced civilizations would be less than thirty light-years apart and could easily detect each other by using radio telescopes. Probes do make sense if advanced civilizations are rare enough so that the distance between them is more than thirty light-years. If L, the lifetime of an advanced civilization, is very large, it is quite possible that an interstellar probe is on the way to the solar system or has already arrived. "One might be in our system now, and if this is the case, we should be very careful not to overlook unexplained radio signals that may be received," Bracewell said in a 1962 lecture.

AN INTERSTELLAR SOS?

One curious episode in SETI is based on the theory that signals from an interstellar probe have already been received on earth. The story, which has gotten much more attention in Great Britain than in the United States, began in 1927, when a Dutch engineer named Balthasar Van der Pol began picking up strange echoes from an experimental shortwave radio transmitter at Eindhoven. Word of the echoes found their way to a Swedish scientist, Carl Störmer, who arranged for a series of experiments using the transmitter. The original echoes had come back with a three-second delay and were interpreted as reflections from the moon. Starting in October 1928, a station in Oslo began picking up echoes with time delays varying from three to fifteen seconds. Thirteen days later, another series of echoes, with time delays ranging from three to thirty seconds, were received.

The matter was allowed to drop until the early 1970s, when it was picked up by Duncan Lunan, a Scottish writer and SETI enthusiast. Lunan has written a book describing his interpretation of the signals. In it Lunan says, "Assuming for the sake of argument that the echo patterns came from a space probe, therefore, I asked, 'What meaning could these signals be meant to convey?' "

Strongly influenced by Bracewell's theories, Lunan was able to establish to his own satisfaction that the time delays in the signals did indeed bear a message, and one of impressive complexity. In 1973 he announced that the time sequence of the signals could be interpreted as a star map of the constellation Boötis, which includes the bright star Arcturus. The fact that Arcturus was slightly misplaced in the star map only added weight to the interpretation, Lunan said. He calculated that Arcturus had been in that position several thousand years ago, when a space probe might have been launched by an extraterrestrial civilization.

In his writings and lectures, Lunan says that his study of the original LDEs (long-delayed echoes) and of others that have occurred more recently indicates that earth has received a message from the star system Epsilon Boötis. As Lunan reads the message, it tells us that there are seven planets orbiting Epsilon Boötis, that the senders of the message originally lived on the second planet and that they have had to move to the sixth planet because their star is growing old and is starting to expand. In effect, Lunan said, earth has received an interstellar SOS message from an advanced civilization in its death throes.

"Travelers" are more common in British SETI circles than in the United States (members of the British Interplanetary Society have a fondness for drawing up plans of hypothetical starships), but Lunan's interpretation of the LDEs seemed to many listeners to be something best approached cautiously. A British electronics engineer, Anthony Lawton, thereupon set about an investigation of LDEs. Lawton not only began transmitting signals to re-create LDEs but also went back to the original Van der Pol work. Lawton's experiments and analysis, done with the help of a friend, Sidney Newton, have convinced him and most other scientists that LDEs are of natural origin.

Using his own antenna, Lawton has been able to get long-delayed echoes with delay times of a few seconds. He believes that these and the longer echoes are due to reflection of the original signals from the ionosphere. When a radio wave enters the ionosphere, Lawton explains, it can be both delayed and amplified. This means that a radio wave comes out of the ionosphere a few seconds later than it is expected to and with greater strength. The ionosphere consists of bands of charged particles. The bands change constantly in size and position under the influence of natural factors. Ionospheric bands of different sizes and positions can explain all long-delayed echoes, Lawton says.

The clincher comes from a study of the original 1927 data, which

Lawton obtained from R. L. A. Appleton, who did some work on the signals at the time. Precise measurements of the signals showed that the echoes had not come in exactly even-second intervals, as might be expected if they had been relayed by a space probe. Instead, the intervals worked out to be imprecise timings: 5.1 seconds, 2.3 seconds, and the like. Instead of the crisp, precise transmissions we would expect if aliens were trying to communicate with us, the long-delayed echoes are the smeared-out, blurred signals that are typical of natural phenomena. Lawton says that he still believes that an alien probe might be somewhere in the solar system, but he does not believe that long-delayed echoes are signals from such a probe. Indeed, in a book he wrote with journalist Jack Stonely, Lawton concluded that "long-delayed echoes would be a most cumbersome and unnecessarily confusing way of making contact. Surely the obvious thing a probe would be programmed to do would be to send its *own* signals and make itself as conspicuous as possible."

The episode of the long-delayed echoes is a notable illustration of the persistent intrusion into SETI of themes that once were the monopoly of science fiction. The willingness to entertain ideas that are beyond the current reach of science and technology is an essential part of SETI; the whole field can be described as a science-fiction idea that has gained scientific respectability. Generally, it is the travelers rather than the listeners in SETI who are most open to speculative ideas. Listeners are in the happy position of championing programs that require little or no advance over current technology and that can be described as cost-conscious. Only the duration and ultimate spending required for listening programs can make a congressman unhappy. The missions envisioned by travelers, however, require an immediate leap to distant technological advances. As the Pioneer 10 mission shows, today's chemical rockets will not get a traveler very far.

But some travelers argue that the interstellar probe is, over the long run, a more sensible undertaking for SETI than a prolonged effort to detect radio signals from another civilization. Bracewell maintains that unless technical civilizations are unusually common in the galaxy, listening programs are not likely to succeed because of many factors: funding bodies on any planet will not be willing to pay for a program that could run for centuries; the chance that any one civilization will be sending during the brief period that another civilization is listening is low; and even if one civilization listens at the proper time, the other planet may be broadcasting on a different wavelength, or may be using a means of transmission beyond the capabilities of the first civilization. "If one hundred light-years is the distance to the next inhabited community, trying to reach them by radio will present extraordinary difficulties and may be doomed to failure," Bracewell said in a 1960 lecture.

TRAVELERS VS. LISTENERS

Some travelers maintain that sending interstellar probes and trying to detect messages from them is a more practical technique than listening for radio signals from another planet. In a 1977 paper published by the journal *Science,* T. B. H. Kuiper of the Jet Propulsion Laboratory and M. Morris of Cal Tech started with the assumption that any fairly advanced technical civilization will be able to achieve interstellar travel at one-tenth the speed of light by using nuclear rockets. "No fundamentally new physics is being advocated," they wrote, "and since no known physical principle states that a deuterium fusion rocket is impossible or unfeasible (it appears to be primarily an engineering task), there is every reason to expect its realization within the next century or so."

Kuiper and Morris also maintained that the distances between habitable planets is not "prohibitively large." They mention not only the space ark ("space ships of sufficient sophistication to permit several generations to live in circumstances not significantly less comfortable than those encountered in crowded cities") but also suspended animation and the possibility that there are "intelligent beings having significantly longer lifetimes than ours." (Other scientists have noted that the distance between habitable planets might be much less toward the center of the galaxy, where the average distance between stars is not much more than one light-year, than it is in the earth's neighborhood, a galactic suburb where stars are five or six light-years apart, on average.)

Since interstellar travel is inevitable if an advanced society has a long life, Kuiper and Morris say there are two possibilities to deal with: case (i), in which technological societies that live long enough to colonize space are rare, and case (ii), in which there are several such civilizations. Either way, they say, the likelihood of making contact with a technological civilization by radio is so low that a long-term listening program is not advisable.

If case (i) is true, they say, then "the galaxy is essentially unpopulated and the number of radio contact beacons is negligible." If almost no one is sending, it would be a waste of time to listen. If case (ii) is true, and one or more advanced civilizations have begun colonizing the galaxy using spacecraft traveling at one-tenth the speed of light, the argument against listening is more complex but no less convincing, Kuiper and Morris say.

Assume that an advanced civilization moves to nearby stars in jumps of 10 light-years, and that the civilization takes 500 years to consolidate each foothold before making the next journey of 10 light-years. The effective rate of expansion is then great enough so that the civilization will colonize the entire galaxy in no more than 50 million years, Kuiper and Morris calculate. There are three implications of this theory, they say:

(1) A "galactic community" would exist, in which one or several civilizations communicate with each other throughout the galaxy, and we are located within the "sphere of influence" of one or more of these civilizations.
(2) The solar system probably would have been visited.
(3) An advanced civilization would probably have representatives somewhere in the solar neighborhood. This encompasses Bracewell's suggestion that an advanced civilization may already have sent an unmanned probe into our system to contact us when we reach some developmental threshold.

If case (i) is true, listening makes no sense because no one is out there. If case (ii) is true, a large-scale listening program makes no sense because any technical civilization is either so close to earth that it is easily detectable using current antennas, or such a civilization is using "beacons of a more sophisticated nature (beyond the capability of our technology to detect)," or because the other civilization already knows we exist and is biding its time—presumably until humans become mature enough to join a truly advanced galactic community. Kuiper and Morris leave open the possibility that UFOs really are extraterrestrial visitors that are monitoring our progress and are making contacts with humans in an unnoticeable way. And, apologizing for venturing "far beyond the realm proper to physics and astronomy," they suggest that the earth has not been colonized by a galactic civilization for several truly cosmic reasons: either our planet is being maintained as a wildlife preserve, or the other civilization has advanced so far technologically that it no longer needs a planetary base, or the biology of earth "is incompatible with or even hostile to that of the species which dominate our part of the galaxy."

LISTENERS VS. TRAVELERS

Naturally, the listeners in SETI are not convinced by the arguments of the travelers. John Billingham, a leader among the listeners, pointed out not long after the Kuiper and Morris paper was published that their argument could easily be used to show that interstellar travel is improbable even for advanced civilizations.

Kuiper and Morris are quite right in saying that interstellar travel at one-tenth the speed of light would not be difficult to achieve, even with only a slight advance over existing terrestrial technology, Billingham says. They are also quite right in saying that a civilization using such a relatively modest (by interstellar standards) mode of travel could spread through the galaxy in a relatively brief period of time (again by interstellar standards). And since the galaxy has been in existence for ten billion years or so, it is also quite probable that a large number of civilizations could have evolved to the point where that kind of space colonization became possible, Billingham says. Nevertheless, he says, we on earth see no evidence for the existence of such a galactic civilization.

To Billingham, it is an unavoidable conclusion that no such colonizing galactic civilization exists. If it did, we would know of its existence by now. Therefore it is logical to assume that interstellar travel is not a commonplace occurrence in the galaxy, even after ten billion years of evolution. Since we must assume that many advanced civilizations have existed in the course of those ten billion years, the absence of a galactic colonizing civilization based on interstellar travel indicates that starships will always be rare phenomena, Billingham concludes.

Obviously, this sort of listener-traveler debate could go round and round endlessly. Equally obviously, the listeners have the best of it now for purely practical reasons. Interstellar travel would cost a great deal of money, even at the most modest level imaginable. That kind of money simply is not available. After the United States spent its $25 billion on Project Apollo, there was a quiet backlash against similar space projects. Spiro Agnew, then vice-president and typically one beat behind, was laughed at when he said the United States should start planning a manned mission to Mars.

THE HIGH COST OF TRAVEL

The thought of such a project, which would have cost an estimated $100 billion in the less-inflated dollars of the 1960s, was so alarming that Congress rather quickly canceled funding for Nerva, a nuclear rocket that would make a manned Mars mission possible. The project was killed even though the nuclear rocket was meeting most of its technical targets on schedule and even though NASA insisted that Nerva would never be used for a trip to Mars. Congress made it clear that cancellation of the nuclear rocket was an insurance policy against a future Mars mission.

For the time being, at least, Congress and the administration still are thinking small about space—"small" meaning a space budget of about $4 billion a year. By their nature, plans for interstellar travel require thinking big. For example, one proposal for a laser-propelled trip to another star calls for unfurling a sail that would be attached to the starship by cables no less than twenty miles long. Dyson's hydrogen-bomb ship would require 30,000 bombs. Enzmann's "frozen iceberg" calls for many tons of frozen deuterium. These are not cheap ideas.

In addition to their costs, plans for interstellar spaceships have another drawback: most of them require technology that has not yet been developed. Indeed, several of them require technology of a sort that has not yet been conceived. Travelers thus must think not only big but also long, in terms of at least decades and most likely centuries before their plans can be fulfilled. An early paper on interstellar travel by Carl Sagan summed the field up best. Sagan wrote that "interstellar spaceflight at relativistic velocities to the furthest reaches of our galaxy is a feasible objective for humanity." But it is feasible, he said, only given "a modicum of scientific and technological progress within the next few centuries."

That time scale should be enough to discourage proposals for interstellar travel now. But there is also another reason why such plans, attractive as they appear on paper, are being discouraged. Interstellar travel now is as much in the realm of science fiction as it is in the area of scientific fact. And the search for extraterrestrial intelligence has begun to reach a state of maturity in which science-fiction proposals are an embarrassment. Spokesmen for SETI are presenting carefully drawn plans for their projects to cost-conscious administrators and legislators. Asking for a few million real dollars is far different from drawing up unrealizable billion-dollar projects. In fact, the existence of those unusual projects could endanger the funding for realizable SETI programs. It is significant that the most imaginative schemes for interstellar travel originate with the British, whose space program is minimal when compared to that of the United States. When nothing is possible, anything is imaginable.

Interstellar travel projects are great fun to write about and marvelous stuff to read. One of them might come to fruition after a while. But for those in SETI who are working on more than fun and games right now, listening is the only game in town. Even Kuiper and Morris concluded their paper by recommending a radio-wave frequency at which listening should be done. Listening will be the primary technique of SETI for the years ahead, as it has been since the field was born.

CHAPTER NINE

LISTENING

In a report that was issued in 1971 to propose a major SETI program called Project Cyclops, Bernard M. Oliver of the Hewlett-Packard Corporation, one of the leaders in the field, summed up the desirable properties of any particles that could be sent the very long distances needed for interstellar communication:

(1) The energy . . . should be minimized, other things being equal.
(2) The velocity should be as high as possible.
(3) The particle should be easy to generate, launch, and capture.
(4) The particles should not be appreciably absorbed or deflected by the interstellar medium.

All these requirements are met by radio waves. Like visible light and all other kinds of electromagnetic radiation, radio waves can be regarded as consisting of particles called photons. The advantages of photons for interstellar communication are obvious, Oliver went on:

Of all known particles, photons are the fastest, easiest to generate in large numbers and to focus and capture. Low-

> frequency photons are affected very little by the interstellar medium and their energy is very small compared with all other bullets. . . . Almost certainly electromagnetic waves of some frequency are the best means of interstellar communication—and our *only* hope at the present time.

That last sentence expresses the opinion of almost everybody in the SETI community now. In the abstract, some travelers believe that the earth ultimately will have to send space probes to other stars. But even the more optimistic travelers are prepared to wait decades before the first interstellar probes are launched. To launch such probes, we would need not only billions of dollars, but also some major technological advances. In the world of here and now, no one is disputing the statement that radio waves are "our only hope at the present time."

"OUR ONLY HOPE AT THE PRESENT TIME"

Put in such terms, the traveling-vs.-listening debate is easy to resolve. Either earth attempts to send, as Oliver says, "tons of metal" over distances of many light-years, or we transmit (or receive) information in the form of radio waves. Since earth already is well equipped with antennas that can send or receive signals from an appreciable portion of the galaxy, and since no great advance in technology is needed to build antenna systems that would measurably increase the chances of detecting signals from another planet, why not start listening and sending at once?

There are a number of factors that complicate the answer to that apparently simple question. One of the most important of them is the choice of a frequency at which to listen or transmit. (As we said earlier, electromagnetic radiation can be described either in terms of its frequency or of its wavelength, which are essentially two ways of describing the same thing; high frequency means low wavelength and vice versa.) The universe is awash with electromagnetic radiation of all frequencies. Planets emit infrared radiation, stars twinkle in visible light, galaxies emit radio waves, other stellar bodies emit X rays and gamma waves, and interstellar molecules emit their own distinctive radiation. If you were an advanced technical civilization whose capabilities can only be guessed at by earthlings, what frequency would you use for your own interstellar beacons and for signals aimed at other civilizations?

The first major attempt to grapple with the question was made by Philip Morrison and Giuseppe Cocconi, who were then physicists at Cornell University, in one of the most important papers in the history of SETI. The paper was published in the prestigious British science journal *Nature* in September 1959, and it played a major role not only in getting SETI started but also in shaping all future discussion of the listening strategy that earth should follow.

To enter this discussion, we need a little vocabulary and an analogy. The vocabulary is that which scientists use to describe radio waves. If we describe the wavelength, the description is in terms of length. Most scientists use what Americans call the metric system, whose unit of length is the meter. Radio waves are best described in centimeters. One centimeter is one-hundredth of a meter and is 0.4 inch long—that is, 2.5 centimeters are roughly equal to one inch. Alternately, electromagnetic radiation can be described by its frequency. The unit of frequency is the Hertz, abbreviated Hz, which refers to cycles per second. Radio waves are in the range of millions of cycles per second, abbreviated MHz (for megahertz).

When Morrison and Cocconi were struck by the idea that interstellar communication by radio waves might be possible—and it was a daring idea for that time—radio astronomers had just discovered the distinctive emission of hydrogen atoms in space. The discovery was just the beginning of the continuing study of interstellar molecules, and it was enormously exciting in the budding field of radio astronomy. The radiation emitted by hydrogen atoms consists of 21-centimeter waves whose frequency is 1420 MHz. Its existence seemed to be an answer to a question that Morrison and Cocconi had been discussing.

THE MAIN STREET OF SPACE

The question is best posed in the form of an analogy. Say that you have agreed to meet a friend in a strange town, and he casually says that you can get together with him on "the main street." If the town is small enough, there is no problem. But what is the "main street" in a city the size of New York, or Boston, or Los Angeles? If you are in New York, should you go to Fifth Avenue or Broadway or Main Street? The question can be answered if the two friends know each other's interests. If they are bankers, Wall Street would seem the logical place to try. If the theater is their main interest, Times Square is the location to explore. If fashion is of interest, try Fifth Avenue.

Morrison and Cocconi assumed that the main interest of any technical civilization that attempts interstellar communication would be science. And therefore, they said in the *Nature* paper, "Just in the most favored radio region there lies a unique, objective standard of frequency, which must be known to every observer in the universe: the outstanding radio emission line at 1420 Mc/sec." (In those days, frequency was expressed in megacycles, or Mc, per second.)

As Morrison and Cocconi noted in their reference to "the most favored radio region," the hydrogen line has one major advantage. It lies in a quiet part of the radio spectrum, as seen from the surface of earth. To ground-based observers, most of the radio spectrum is quite noisy, so much so that it is impossible to detect individual sources of specific frequencies in space. The noise is caused by molecules in the earth's atmosphere—molecules of water and carbon dioxide and ozone and all the other gases in the air. These molecules absorb specific frequencies of incoming radiation. The molecules then emit radiation of the same frequencies, in effect blocking the extraterrestrial signals. Over a large portion of the radio spectrum, a radio astronomer on earth receives nothing but emissions from the atmosphere, which are no good at all for anyone who wants to study emissions from space.

But there is a "window": a region of relatively low atmospheric noise, that starts at about 1,000 MHz and extends to about 10,000 MHz. This is the "most favored radio region" to which Morrison and Cocconi referred. The hydrogen emission line of 1,420 MHz is in this window. So is the emission line of the hydroxyl ion, which is made of one hydrogen atom joined to one oxygen atom, and which emits at 1,662 MHz. Between these two emission lines is the quietest part of the radio spectrum as seen from the surface of the earth.

One other point: if a hydrogen atom is joined to a hydroxyl ion, the product is a water molecule: H_2O. And so Bernard Oliver says, "The Cyclops team feels that this band, lying between the resonances of the two dissociation products of water, is the foreordained interstellar communication band. What more poetic place could there be for water-based life to seek its own kind than the age-old meeting place for all species: *the water hole?*"

THE "WATER HOLE" OF SPACE

Oliver, who has a tendency to wax poetic, expanded on the thought in a 1977 report of a SETI committee headed by Morrison:

> It is easy to dismiss this as romantic, chauvinistic nonsense, but is it? We suggest that it is chauvinistic and romantic but that it may not be nonsense.
>
> It is certainly chauvinistic to water-based life, but how restrictive is such chauvinism? Water is certain to be outgassed from the crusts of all terrestrial planets that have appreciable vulcanism and, therefore, a primitive atmosphere capable of producing the chemical precursors to life. We can expect seas to be a common feature of habitable planets. Exobiologists are becoming increasingly disenchanted with ammonia and silicon chemistries as bases for life. Water-based life is certainly the most common form and well may be the only (naturally occurring) form.
>
> Romantic? Certainly. But is not romance itself a quality peculiar to intelligence? Should we not expect advanced beings elsewhere to show such perceptions? By the dead reckoning of physics we have narrowed all the decades of the electromagnetic spectrum down to a single octave where conditions are best for interstellar contact. There, right in the middle, stand two signposts that taken together symbolize the medium in which all life we know began. Is it sensible not to heed such signposts? To say, in effect: I do not trust your message, it is too good to be true.

In the absence of any more cogent reason to prefer another frequency band, we suggest that the water hole be considered the primary preferred frequency band for interstellar search.

In fact, the water hole is today the most favored region of the radio spectrum for SETI. All plans for detecting signals from another civilization center on radio waves, and almost all of those plans are limited to the water hole. However, some thought has been given to other possible forms of interstellar communication.

One conceivable way of sending messages over interstellar distances is the laser, which emits an intense beam of light of very narrow wavelength. As early as 1961, Charles Townes, who was awarded the Nobel Prize in physics for developing a forerunner of the laser called the maser (the "m" in maser stands for "microwave") suggested that the earth could wink at other civilizations across interstellar distances by reflecting the beam of an extremely powerful laser off a very large mirror. The laser could transmit a message by blinking off and on, in a sort of interstellar semaphore.

It would seem at first that the light from any laser would be weak compared to that of the sun, but this problem can be overcome rather easily. There are a number of dark lines in the sun's emission spectrum, lines which mark the presence of elements that absorb light of specific wavelengths. Lasers that emitted light at one or more of these wavelengths could be used for the interstellar semaphore. Astronomers on another planet could detect the signal by noting first that the spectrum of the sun did not have a dark line at a frequency where one is expected; that, in fact, the sun was emitting an unusually large amount of energy in that frequency; and finally that the emissions were pulsed in such a way as to carry a message. It is true that the laser signal would be weakened by atmospheric absorption but that problem could be eliminated by putting the laser system into orbit around earth.

At the meeting in Armenia, Townes argued for a bigger SETI emphasis on lasers, which he said are especially suitable for communication with relatively close stellar neighbors, within ten light-years or so. "Within five years our own possibilities will change and any civilization that may be one hundred years beyond us may see things quite differently," he said. A civilization that had unlimited cheap energy and that lived on a planet with unusually intense winds might well prefer to use laser interstellar beacons, which require more power but smaller (and therefore less vulnerable) antennas than radio transmitters, Townes said.

In theory, lasers are a viable tool for SETI. At the moment, however, no one is talking with great enthusiasm about interstellar laser beacons. Money is one reason. Any interstellar communication system using lasers would be extremely expensive and would have no other use. The money to build such a system simply is not available. In addition, SETI is now concentrating on detecting signals from another civilization, not on sending interstellar signals. Radio waves can be detected with existing radio telescopes, which can be used for astronomical observations when they are not needed for SETI. To detect light signals from another civilization, optical telescopes must be used. There is no surplus of optical telescopes, and viewing time on them is a scarce commodity. There have been no recent suggestions for a galactic search to detect intelligent signals in the optical region of the electromagnetic spectrum.

Some scientists in the SETI community entertain the thought that the water hole, which is an obvious galactic meeting place for a society in our stage of technical development, may be too primitive for a more advanced society. As Carl Sagan said in Armenia:

> Such societies will have discovered laws of nature and invented technologies whose applications will appear to us indistinguishable from magic. There is a serious question about whether such societies are concerned with communicating with us, any more than we are concerned with communicating with our protozoan or bacterial forebears. We may study microorganisms, but we do not communicate with them. . . . We may be like the inhabitants of the valleys of New Guinea who may communicate by runner or drum, but who are ignorant of the vast international radio and cable traffic passing over, around and through them.

SIGNALS TO EXPECT

There are a few theories around about the sort of signals we might expect from very advanced societies. Nikolai S. Kardashev, a Soviet radio astronomer, has classified civilizations by their technological stage of development. In Kardashev's grouping, a civilization with the power output now possible on earth would be in Type I. A Type II civilization would be able to harness the total energy output of a star like the sun; at the earth's present rate of energy growth, our civilization would become Type II in about 3,000 years. Finally, there is a Type III Kardashev civilization, capable of using energy equivalent to the output of an entire galaxy.

In the United States, Freeman Dyson has proposed a comparable vision of an advanced technological society. A "Dyson civilization" would have the ability to literally take apart a planet and reassemble its mass in a shell around a star, thus trapping all or most of the star's radiation. Originally, Dyson thought in terms of reshaping the planet into a solid shell around the star. The mass of a planet the size of Jupiter would form a shell a few yards thick with a radius of a little more than 90 million miles, about the distance of the earth from the sun. People could live comfortably inside such a shell, which would give mankind the use of all the sun's energy, rather than the almost negligible fraction that now supports life on earth.

However, later calculations indicated that a Dyson sphere would be gravitationally unstable, so now a Dyson civilization is envisioned as using a swarm of planetoids to trap the energy of its star. Such a scheme would still give a civilization the use of an incredible amount of energy by current earthbound standards.

Both the Kardashev and Dyson proposals lend themselves to observational testing. A Dyson civilization would change the radiation output of a star quite drastically, as seen from a distance. Instead of radiating primarily in the visual and radio regions of the spectrum, a star surrounded by a Dyson sphere would emit primarily infrared radiation; the planetoids would absorb the higher-frequency radiation and would re-emit it in the form of infrared radiation, or heat. A Kardashev Type III civilization would have enough energy at its disposal to establish interstellar beacons whose distinctive signals could be detected through a large part of the universe. Kardashev went on to say that some known radio wave sources in the universe could possibly be such universal beacons. He named two in particular, cosmic radio sources named CTA-102 and CTA-21, whose signals have been seen to vary rhythmically. The variations could be carrying a coherent message, Kardashev said.

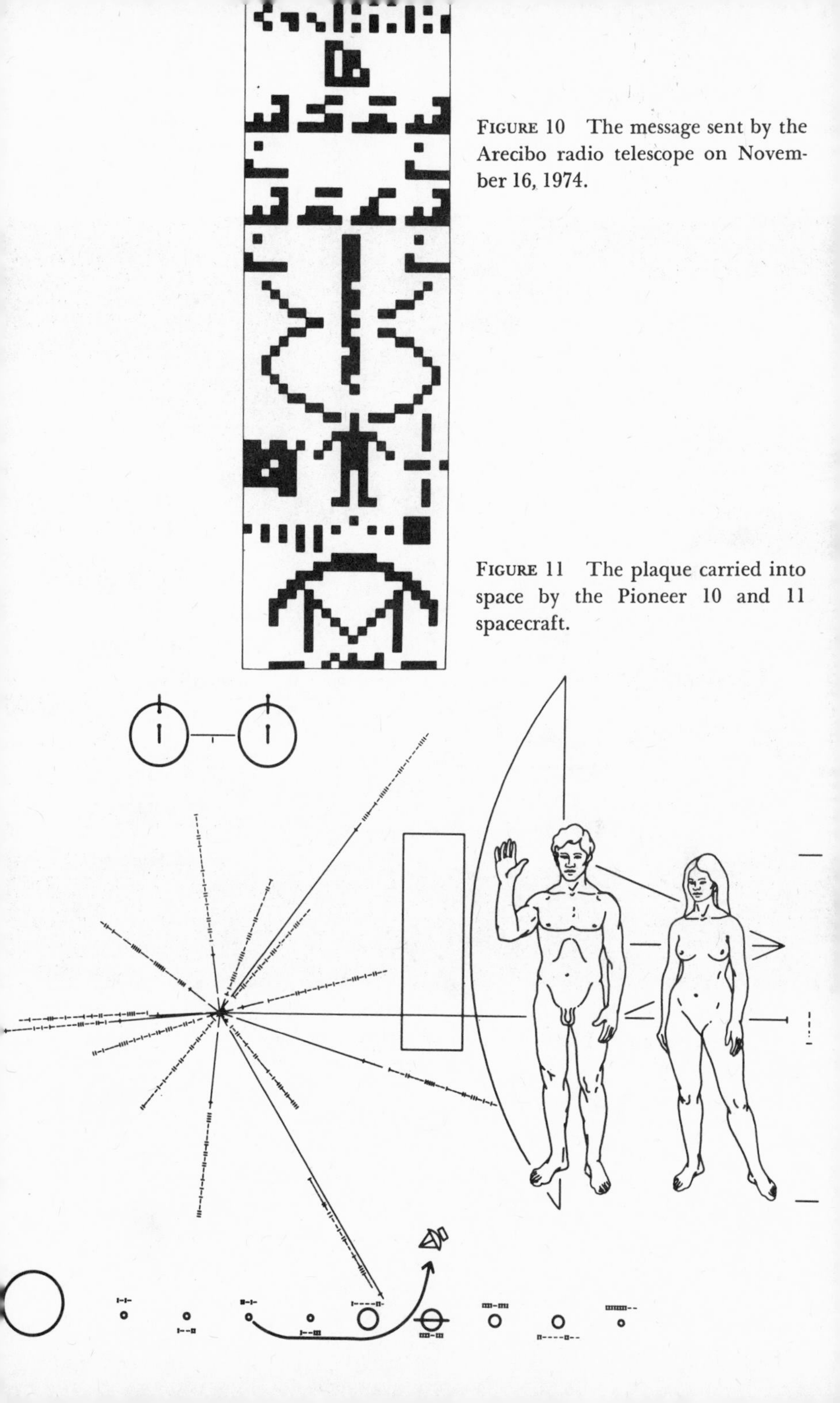

FIGURE 10 The message sent by the Arecibo radio telescope on November 16, 1974.

FIGURE 11 The plaque carried into space by the Pioneer 10 and 11 spacecraft.

FIGURE 12 Part of the orchard of antennas of Project Cyclops, as seen from the ground. Each antenna is 330 feet in diameter.

FIGURE 13 Cyclops as seen from the air.

FIGURE 14 An artist's concept of three Arecibo-type antennas on the moon.

FIGURE 15 An artist's concept of a relatively small SETI system in space. The 1,000-foot-diameter antenna is shielded from radio interference by the disc between it and the earth.

FIGURE 16 The two radio antennas, one 85 feet in diameter, the other 35 feet, at Goldstone, California, that will be used in the Jet Propulsion Laboratory's listening program.

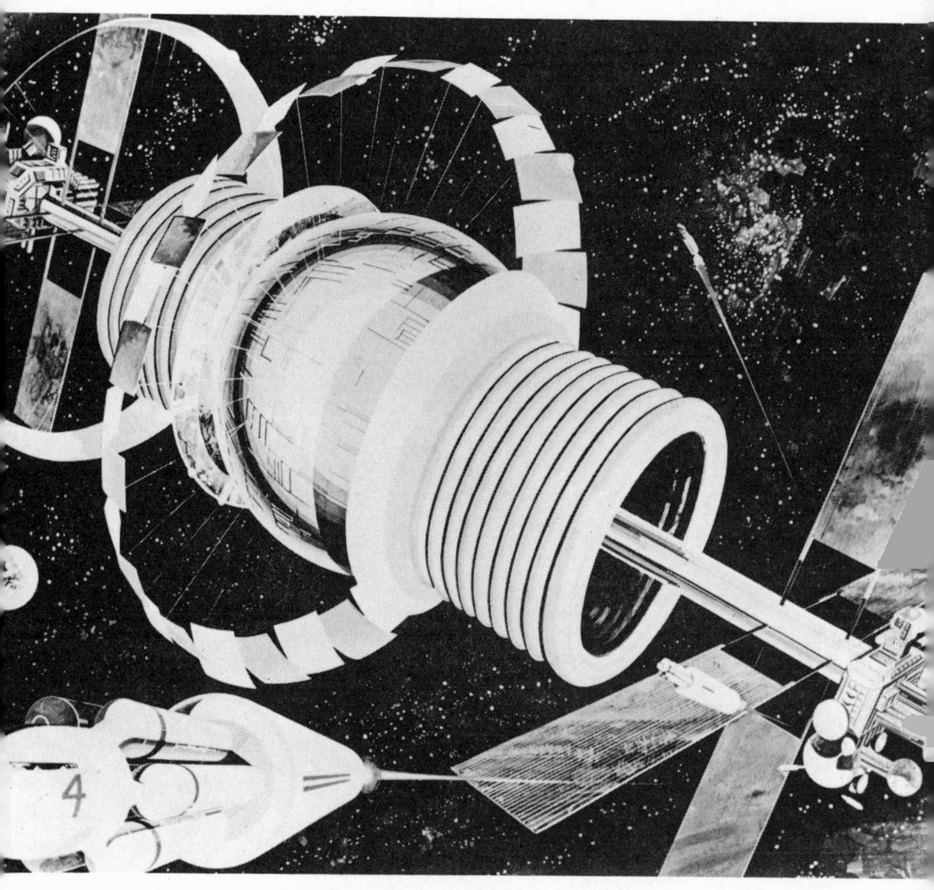

FIGURE 17 A Bernal sphere that could house 10,000 people in a space colony, as envisioned by Gerard K. O'Neill.

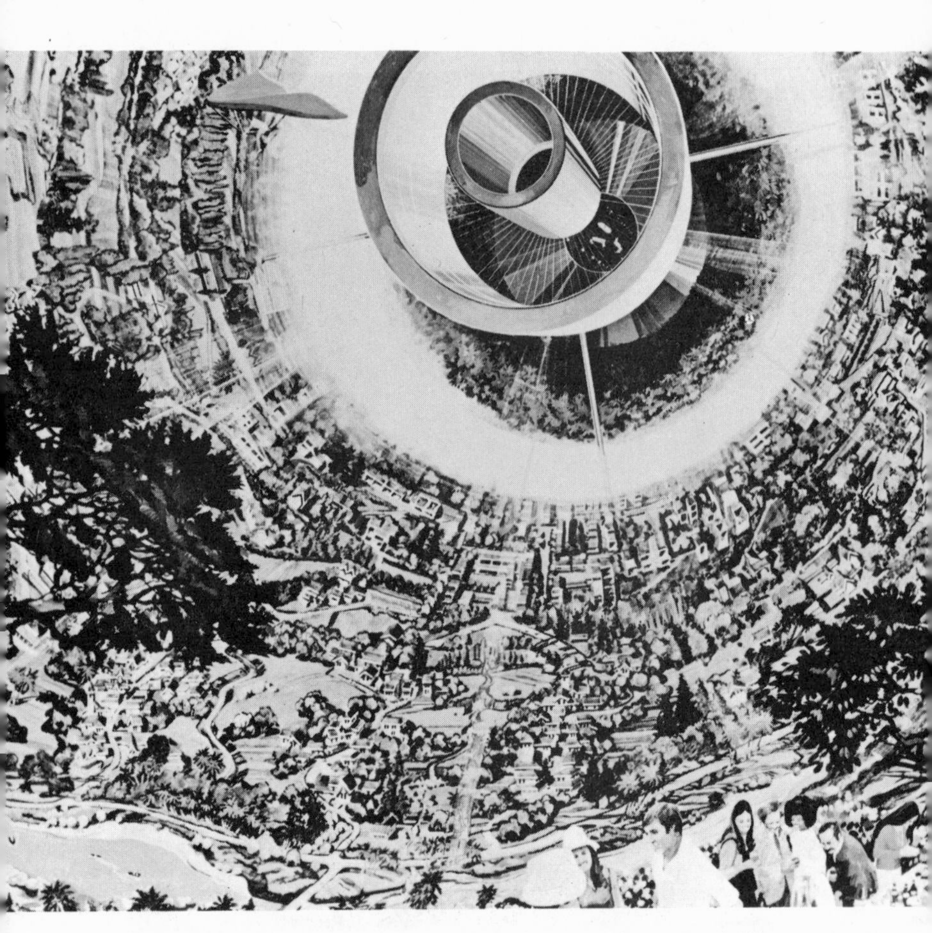

FIGURE 18 The interior of a proposed space colony. Earth-like gravitational pull is provided by spinning the colony slowly. The corridor in the center has zero gravity and leads to industrial areas, farms, and other facilities.

Astronomers have studied both CTA-21 and CTA-102 to determine whether their signals do carry a message. Several stars and galaxies that emit radiation primarily in the infrared have also been studied to test the Dyson hypothesis. In both cases, the conclusion has been the same: if these objects are the work of intelligent beings, current science cannot say so. The peculiarities in the radiation emissions look natural, not artificial. If a Kardashev Type III civilization or a Dyson civilization does exist and if it is trying to communicate across space, its signals are too sophisticated for earth to detect now.

A few wilder speculations float around in SETI. Some scientists toy with the idea that an advanced civilization communicates by using tachyons, which are hypothetical particles that can travel *only* faster than light. However, such speculations do not get SETI very far on the practical level. The tachyon hypothesis is fascinating on the theoretical level—for example, a tachyon message would arrive at its destination before it left its point of origin—but no one has any idea of whether tachyons actually exist or how to harness them if they do exist.

Imaginative thinkers in the SETI community are constantly on the alert for some new development in physics that could provide an advanced alternative to communicating with radio waves. Nothing done on the frontiers of physics in the past couple of decades has produced such an alternative. In fact, the most detailed studies done recently have hardened support not only for communication by radio waves but also for using the water hole for interstellar signaling.

SETI: FROM ENFANT TERRIBLE TO MIDDLE AGE

There are some practical reasons why the options for interstellar communication appear to have narrowed. The SETI community is still rather small, and a tendency for inbreeding has become apparent. The same names appear over and over on committees and panels. From a start as a scientific *enfant terrible,* SETI is beginning to settle into a comfortable middle age. In addition, radio-wave communication has the great advantage of being cheap and of being possible with existing equipment, extremely important factors as SETI begins to seek government funding. But there are also good scientific reasons: no really practical alternative to radio waves has developed, and the water hole in the radio spectrum seems more likely because of the biological work which indicates that life elsewhere in the universe probably is based on water, as it is on earth.

But even if we agree not only on radio waves but also on the water hole, the next question is: what strategy do we follow? Interestingly, the answer that is most accepted today is essentially the same one given in the very first days when interstellar signaling began to seem feasible.

Even before the Morrison-Cocconi paper was published in *Nature,* Cocconi wrote a letter to Sir Bernard Lovell, director of the Jodrell Bank radio telescope in England, at that time the largest radio antenna in existence, suggesting a search for "beams of electromagnetic radiation modulated in a rational way, e.g. in trains corresponding to the prime numbers." The tone of Cocconi's letter was half-apologetic for proposing a project that "at first sight . . . looks like science fiction," and Lovell replied, briefly, that the observatory was too busy with other things to undertake such a search.

PROJECT OZMA

But the same idea that had occurred to Morrison and Cocconi had also struck Frank D. Drake, a young radio astronomer who was then at the new National Radio Astronomy Observatory at Green Bank, West Virginia. Even before Cocconi wrote to Lovell, Drake had conceived what he called Project Ozma, named after a princess in the L. Frank Baum stories of the imaginary Land of Oz, to search for intelligent signals from space. Drake was fortunate enough to have the support of the observatory's director, Otto Struve, who made time available for Project Ozma despite strong criticism and even ridicule from a large number of astronomers.

Drake chose two nearby sunlike stars, Tau Ceti and Epsilon Eridani, for Project Ozma. The program began in April 1960. Listening was done in the neighborhood of the hydrogen line, the

21-centimeter or 1,420 MHz region. Drake's team periodically tuned the equipment to monitor about 4,000 specific frequencies, each for a minute or so, from each of the two stars. Project Ozma went on for two months, a total of about 150 hours of actual listening, before the Green Bank radio telescope went back to its normal observations.

Project Ozma is important not only as a historic first, but also because it is the prototype for almost every listening project in SETI, actual or proposed, ever since. While one or two searches for extraterrestrial signals have taken a slightly different approach, most listening programs have differed from Project Ozma only in detail. The changes that have occurred in listening programs have been due primarily to advances in technology. Considering that both astronomy and electronics have made major advances since Project Ozma, the continuing reliance on its basic plan is a tribute to Drake's judgment.

Even though the importance of the hydroxyl emission line in space had not been discovered at the time of Project Ozma, its search was carried out in what would later become known as the water hole. In the two months of listening, a large number of what were then regarded as relatively narrow bandwidths, each about 100 cycles wide, were monitored, each for a brief period. Narrow-bandwidth monitoring is a basic principle of most listening programs, on the theory that any signal from an advanced technological society will be narrowly focused. By contrast, natural radio sources have wide bandwidths; they usually are smeared out over many thousands of cycles. But as everyone who has ever tuned in a radio program knows, artificial signals are concentrated in a narrow bandwidth. Even a slight turn of the radio dial is enough to lose a station.

One of the important changes in listening programs since Project Ozma is the tendency to listen to narrower and narrower bandwidths. A number of theorists believe that a signal from another civilization will be detected only by listening to a large number of extremely narrow bandwidths. These theorists believe that an extraterrestrial civilization's signals are likely to be only a fraction of one cycle wide. Listening to narrow bandwidths makes the job much more difficult and time consuming, since the number of frequencies that must be scanned is greatly increased. It is only by using the latest electronic processing equipment that such a listening program can be possible in a reasonable length of time. The cost of the processing equipment is not impossibly high, thanks to the great progress made in miniaturizing electronic circuits, but the processor does have to be built specially. In addition, the processor must be used specifically for a listening program, with not much spillover into general astronomical observations. Those last two points have become important in a growing controversy that threatens to divide the SETI community, as we will see in a few pages.

First, two other characteristics of Ozma should be mentioned, since both of them have become SETI commonplaces. The results of the listening program were negative, but there was a moment when it appeared possible that an extraterrestrial signal was being received.

Ozma's negative results were expected. Even in 1960, the available calculations about the possible number of extraterrestrial civilizations indicated that a relatively brief look at only two nearby stars was not likely to put earth in contact with another intelligent species. To this day, SETI is unusual among scientific disciplines because its practitioners begin their experiments with the expectation that they will not be successful, at least over the short run (and "short" means years). The SETI community thinks in terms of decades or, in pessimistic moments, centuries, as the time required to make contact with an extraterrestrial civilization.

SETI is such a young discipline that, by the law of averages, its results are bound to be negative. If there are one million advanced civilizations in our galaxy, arithmetic shows that only one star in 100,000 has such a civilization. The probability of detecting a signal from one of those civilizations in the first scan of two, or a dozen, or a hundred or even a thousand stars is extremely small. Yet if earth is to detect a signal from another civilization, we must start somewhere. Project Ozma was that start. The fact that it found nothing was less important, in the minds of the SETI community, than the fact that the attempt was made at all.

The false alarm, an apparently coherent signal from outer space that later is found to have a terrestrial explanation, is a less well publicized feature of SETI projects that occurred first in Ozma. Not long after the antenna was pointed at Epsilon Eridani, a strong train of regularly pulsed radio signals was detected for a few minutes. No announcement of the detection was made. Several weeks later, Drake's researchers were able to determine that the signals came from a military facility that was doing radar research.

FALSE ALARMS

Even in the small number of limited SETI projects that have been undertaken since Ozma, false alarms keep occurring. Searchers in the United States and Canada have been extremely cautious about announcing the reception of signals from outer space—a caution whose wisdom has been borne out by experience. The Soviet SETI community has been more willing to rush into print and has been burned several times when announcements about "extraterrestrial signals" have had to be withdrawn.

The most fascinating false alarm of all did not occur in any organized search for intelligent signals but as part of an ordinary exercise in radio astronomy. In 1967 a graduate student named Jocelyn Bell who was part of a group of radio astronomers at Cambridge University in England was doing a tedious analysis of signals that had been received by a new radio telescope when she noticed something odd. A regular, brief blip was showing up on the record—a series of unusually regular pulses at an interval of just over one second. The almost incredible regularity of the signals—they are precise to one part in 100 million—and the fact that analysis showed they had to be coming from a very small source seemed to argue against a natural origin. In the first euphoric hours after the discovery, the Cambridge astronomers named the source LGM-1. The initials stood for "little green men."

The Cambridge radio astronomers did not think that they had received a coherent message from outer space. But the signals did appear to have the characteristics of an interstellar beacon system that a galactic civilization might create to guide its starships. However, the thought that the signals might come from the galactic equivalent of lighthouses faded as other radio observatories swung into action. The LGM designation was dropped quickly. The sources of these precise pulses of radio signals today are called pulsars, and dozens are known to exist. A pulsar is believed to be a neutron star, one of the last stages in the life cycle of stars several times more massive than the sun. When such a star has exhausted its fusion fuel, it flares up briefly in a supernova explosion and then contracts until it has the density of an atomic nucleus. At this density, a star with several times the mass of the sun would be about as large as Manhattan Island. Spinning swiftly, the neutron stars hurls bursts of radiation into space at each revolution. When a burst reaches us, we can detect a pulse of radio waves.

The discovery of pulsars fulfilled a prediction that had been made by astrophysicists about the last stages of the life cycle of stars. But interstellar beacons, no.

The most widely publicized Soviet false alarm occurred in 1973. In October of that year, the Soviet press agency Tass carried a report saying that Soviet radio astronomers had detected radio signals that occurred at regular intervals and did not seem to be of terrestrial origin. Later, one of the astronomers, Nikolai Kardashev, was quoted as saying that the signals "might have come from an artifact which had entered the solar system"—that is, an interstellar probe.

Until now, the Soviet approach to SETI has been somewhat different from that adopted in the United States. U.S. searchers like to use highly sophisticated equipment in an effort to detect subtle signals from individual star systems. The Soviets have tended to use comparatively simple antennas in an effort to detect powerful, unmistakable signals from a highly advanced society, such as a Kardashev Type III civilization.

Starting in 1968, Vsevolod Troitsky and other radio astronomers did make an Ozma-type search of about a dozen nearby sunlike stars using a radio telescope at Gorky State University. They heard no artificial signals at the two frequencies they tried, the 21-centimeter line and the 30-centimeter line. Later, Troitsky's group set up a network of six widely spaced stations—in Siberia, the Crimea, at Gorky, and at Murmansk, and two locations—to detect pulsed signals from the sky at wavelengths of 16, 30, and 50 centimeters. Kardashev and a colleague, Lev Gindilis, established a comparable network using stations in the northern Caucasus and in the Pamir Mountains, about 1,800 miles apart.

Both networks did detect a number of regular pulses that stood out from the background noise. The origin of most of those pulses was tracked down. Some came from an artificial satellite; others from natural emissions from the sun or the ionosphere. The signals that led to the Tass report in 1973 appeared to have no such explanation. They did seem to be coming from a satellite, but not from any known satellite of terrestrial origin. Hence Kardashev's reported belief that a space probe of the sort described by Bracewell had been detected.

Western scientists were skeptical from the start, and their skepticism has been borne out. The latest word from the Soviet Union is that the signals were traced back to a military satellite launched by the United States. An earlier Soviet report in April 1965, in which Tass said, "Scientists at the Shternberg Astronomical Institute believe they have received perhaps the first evidence that we are not alone in the universe," was also a false alarm. The periodically varying radio signals detected by astronomer Gennardy Sholomitsky were found to come from a quasar, a natural (if still somewhat mysterious) body in space. It was a Soviet counterpart of the LGM episode.

One measure of the hold that SETI has on the minds of many people is the failure of false alarms and negative results to discourage interest in the search. At the annual meeting of the American Association for the Advancement of Science in Boston in 1974, one of the biggest conference rooms was filled to capacity for a session on SETI, even though the subject matter was nothing more than unsuccessful searches. The session was a milestone because its arrangers were able to report on an unprecedented total of six searches for extraterrestrial civilizations. Most of them are still going on.

Outside the Soviet Union, the first search after Ozma was made at Green Bank by astronomer Gerritt Verschuur. In 1971 and 1972, Verschuur looked at the hydrogen line of ten nearby stars, including Tau Ceti and Epsilon Eridani. As expected, he heard nothing unusual.

In 1972 Patrick Palmer of the University of Chicago and Ben M. Zuckerman of the University of Maryland began what they called Project Ozma II. The differences between the two Ozma projects show how fast SETI had moved in a dozen years. Instead of just two stars, Palmer and Zuckerman studied hundreds, and their equipment took just a fraction of a second to accomplish what had required 150 hours of listening in 1960.

Project Ozma II used the 300-foot antenna at Green Bank and listened on 384 separate bandwidths centered on the hydrogen line (21 centimeters or 1420 MHz). The stars that were selected for the survey were "very nearby on an astronomical scale; that is, within about sixty light-years of the earth." Palmer and Zuckerman studied only sunlike stars that were prime candidates for life; stars that are too old or too young, those that are too big or too small were eliminated. Even with these restrictions, they had a list of more than 500 stars to scan.

They made their observations in week-long blocks of time between November 1972 and August 1975. During each week, about 150 stars were monitored for about four minutes a day. After using the 300-foot antenna to study the original list of stars, the 140-foot antenna at Green Bank was aimed at about 130 more, bringing the total to 659 stars. In addition, the 140-foot antenna was used for a further scan of ten stars whose original data contained something unusual. In all, Palmer and Zuckerman accumulated nearly 20 million data points in some 51,700 individual star spectra. They first ran the data through the computer and then checked it all again by hand. Most of the "glitches" in the data were explainable. The few that couldn't be traced back to terrestrial sources were one-time phenomena, rather than the repeated trains of pulses to be expected in an interstellar beacon. Results: negative.

Up in Canada, Alan H. Bridle of Queens University in Kingston, Ontario, and Paul A. Feldman of the National Research Council began looking at the water emission line (2,220 MHz), using the 150-foot radio telescope at the Algonquin Park Radio Observatory, in 1974. At the 1976 meeting, they reported that they had scanned 28 stars, and, despite frequent equipment trouble, hoped to examine a total of 500 stars. There were, of course, no intelligent interstellar signals, but astronomers at Algonquin Park were able to detect the heaviest interstellar molecule found until then, a compound called cyanodiacetylene. By early 1978, Bridle and Feldman had looked at 150 stars, still with negative results.

GALACTIC SWEEPS

Other searchers are a bit more off the beaten path. Carl Sagan and Frank Drake have used the 1,000-foot radio telescope at Arecibo to study whole galaxies. The advantage of such a search is that it is a nice, wholesale approach to SETI. Instead of surveying one star at a time, a galactic search takes in anywhere from 100 billion to 250 billion stars in a swoop. The disadvantage is that only a highly advanced civilization that is beaming extremely strong signals can be detected. Drake estimates that the Arecibo telescope, the largest on earth, could pick up signals in a galactic search only if they were sent with a million times the maximum power of the Arecibo antenna. The Arecibo galactic sweep uses four different frequencies—the hydrogen line, the hydroxyl line, a line at 1,652 MHz in the middle of the water hole, and 2,380 MHz, which happens to be the frequency that the Arecibo telescope utilizes when it is used as a radar facility. The first galaxies monitored were Leo 1, Leo 2, Messier 33, and Messier 49. There was the usual false alarm, a moment when a peak appeared in the 1420-MHz signal from Leo 1, but the same peak appeared when the antenna was not pointed at the galaxy. The signal probably came from a gas cloud in interstellar space.

Sagan and Drake hope to do more galactic sweeps (Bridle and Feldman have a list of 150 galaxies they would like to study with the Arecibo radio telescope), but their project is handicapped because the search for extraterrestrial signals must be fitted into unused moments in the telescope's schedule. Those moments are rare since radio astronomers queue up for months to get observing time at Arecibo.

At Ohio State University, radio astronomer Robert S. Dixon and some colleagues are trying a low-cost approach to SETI. They have rigged up the equivalent of a 175-foot antenna using equipment scrounged from various sources, and they run their search almost around the clock, leaving the machine unattended. The Ohio State device is methodically surveying as much of the sky as is possible at the 21-centimeter line. Dixon and his group started the scan in December 1973; as of 1977, they had detected no unexplainable signals in the segment of the sky they had looked at (between $+48°$ and $+14°$ declination, for the record). The few blips that did show up were traced back to terrestrial radar facilities. However, the Ohio State group did pick up signals from a number of natural radio sources that appeared worthy of further astrophysical study.

A couple of points about these searches: negative results for SETI are not necessarily completely negative for astronomy. Useful astronomical data is being collected even in the first rather limited and inexpensive efforts. As the SETI community begins to press its case for a bigger, better-funded program, it is starting to stress the benefits to astronomy of its scans.

And there's a lot more going on in SETI than most people realize. Add up all the observations that have been made, and it seems that many hundreds of stars and a fair fraction of the sky have been surveyed in one way or another for intelligent signals. After hearing the reports at the American Association for the Advancement of Science meeting, Philip Morrison suggested that the time had come for SETI to get organized, so that duplication of effort would be avoided. There are two sides to that question. It is certainly true that, even at this early stage of the search for extraterrestrial signals, searchers tend to repeat what others have already done. For example, almost everyone who has started a search program seems to have looked at Tai Ceti and Epsilon Eridani, Drake's two original targets for Ozma. Barnard's star is another favorite. And, in addition to looking at the same stars, searchers stick to the same frequencies, usually the 21-centimeter line.

On the other hand, repeated searches of the same stars in the same frequencies are not necessarily wasteful. We cannot assume that an advanced civilization will beam an interstellar beacon constantly toward the earth; in fact, it is safer to assume that such a beacon would be directed toward earth only intermittently. Therefore it seems prudent to keep checking likely stars periodically. Since the number of stars that has been scanned so far is not much more than 1,000, SETI cannot be said to have reached a point where duplication of effort is a real problem.

But that point may be reached as more searches are begun. As of 1978, there were eight continuing programs—one in the Soviet Union, the Bridle-Feldman program in Canada, and six in the United States. The last entry on the United States list is an indication of how searches might multiply because of the effects of the cost revolutions in electronics.

The cost of carrying out a search for extraterrestrial signals is going down because the cost of the electronic equipment needed for the search is dropping extremely rapidly. A single chip that now costs less than $20 contains the circuitry that went into the heart of a million-dollar computer of the 1950s. A searcher can now use such low-cost equipment for a low-cost listening program that involves very little effort.

Late in 1977, a group of graduate students working under the direction of Stuart Bower, a professor of astronomy at the University of California at Berkeley, set up such a program. They call it SERENDIP, which is an acronym for Search for Extraterrestrial Radio Emission from Nearby Developed Intelligent Populations. The acronym is a reference to serendipity, the knack of finding useful things even though one is not looking for them.

For about $4,000, the students built a system that is attached to the 85-foot radio telescope at the university's Hat Creek Observatory, near Mount Lassen in northern California. The system consists of a noise analyzer, a computer, and a tape recorder. As the antenna goes about its ordinary business, the analyzer and the computer search 100 different frequencies for unusual signals. Anything out of the ordinary trips the tape recorder into action, recording not only the location of the signal but also the time when it was received. The tape is reviewed once a month at Berkeley to determine whether any of the unusual signals appears to be an intelligent transmission.

Other listening programs are also being done on a sporadic basis, when time can be spared at major installations. For instance, Frank Drake and Mark A. Stull of the Ames Research Center used the Arecibo antenna in 1977 for a study of six stars. They looked at frequencies in the 1,664–1,668 MHz, near the hydroxyl line. Also in 1977, five days of observing time on the 300-foot antenna at Green Bank were allotted to four astronomers: Jeffrey N. Cuzzi, Thomas A. Clark, Jill C. Tarter, and David C. Black. They looked at 200 sunlike stars within the 1,665–1,667 MHz range, again near the hydroxyl line.

Both searches were negative, of course. They do, however, add to the growing volume of what can be called background information in SETI: information that is comparable to the many thousands of photographic plates that astronomers have accumulated over decades of viewing and photographing the stars. When anything new is found in the heavens, astronomers can go back through the old plates to see whether there were previous sightings of the phenomenon. In the same way, astronomers who are searching for unusual signals from advanced civilizations may someday be faced with the problem of analyzing a borderline signal from a star. If the information that is being gathered by a growing number of individuals is properly filed and indexed, it can help greatly in determining whether the signal does indeed contain a message.

Of course, the search for a message in a signal is helped greatly if the searcher has an idea of what form a message might take. Scientists in the SETI community have been thinking about that problem for a long time. It might not seem easy to visualize a message from beings whose nature is unknown and whose home planet is completely foreign to us. But a start has been made at a theory of interstellar signaling.

CHAPTER TEN

MESSAGES

Listening programs assume that someone is sending a message. But what sort of message? The initial answer to that question is: something very like the signals that pulsars transmit. Martin Ryle, who was to share the Nobel Prize in physics for work that included the discovery of pulsars (Jocelyn Bell did not get a share of the prize, which many observers believe is unfair), later recalled that when the discovery was made, "Our first thought was that this was another intelligent race trying to reach us."

But that does not really answer the question. Assuming that an intelligent species is trying to reach out across almost incalculable distances to make contact with another species, what can it say? Remember: the two species have nothing in common but intelligence; their star systems, planets, genetics, environment all are likely to be quite different. Remember also that human beings, who share the same genetics, planet, brain, and body can have great difficulty in understanding each other over distances of only a few miles. The difficulties are compounded if human contact is made over greater distances. Much of the history of the nineteenth and twentieth centuries is the unhappy story of the collision of two human societies that share everything but a common background. If members of a species on a single planet can encounter such problems trying to communicate, what message can be sent from one intelligent species to another across the void of space?

At the beginning of the twentieth century, the idiosyncratic inventor Nikola Tesla, who thought he had detected signals from another planet, had the essential idea of what they should contain. He envisioned the time when mankind received "a message from another world, unknown and remote. It reads: one . . . two . . . three . . ."

Tesla's idea contained the core of the concepts that most people in the SETI community still hold about interstellar messages. The SETI community, which consists almost entirely of scientists and engineers, assumes that a SETI community on another planet will have the same technical orientation. Scientists and engineers talk largely in mathematics when they are in formal communication with one another. Therefore it is assumed that scientists and engineers trying to talk across space will also use mathematics as their language. As Tesla understood, two intelligent groups that have no words in common cannot help but understand the basic principle of counting, starting with a single unit and adding one at a time. The simplest way to make an intelligent species on another planet understand that a coherent message is being sent is to start with a single pulse, follow it with two, follow that with three, and continue up to, say, ten, and then start again.

But simplicity does not mean efficiency. Interstellar communication would take place across such vast distances that the transmitting society would want to reduce the time for a reply to a minimum. If the receiving society were only fifty light-years away—not a very great distance by galactic standards—the senders could not expect an answer in less than a century. If the first society says only the interstellar equivalent of "hello," it will be 150 years before a real dialogue can begin: 50 for the message to be received, 50 for an answer to be transmitted, and another 50 for the first society to reply.

The round-trip time can be cut considerably if the original message includes a large amount of basic information about the transmitting society: its location in the galaxy and in its planetary system, the genetics of the transmitting species, and other organisms on its planet, and so on. At the conference in Armenia, several of the participants toyed with the idea of transmitting a train of signals that could be put together at the receiving end as an actual picture of an earth creature such as a cat. But S. Y. Braude, one of the Soviet scientists at the meeting, pointed out, "You are transmitting a picture, but that same picture will be received in another world and interpreted in a certain way. Don't you think that it may happen that in another world with creatures with a different refractive index, the pictures will be distorted and the cat will be nothing like the cat because the medium is different?"

SENDING MATHEMATICAL PICTURES

The current reigning concept in the SETI community nevertheless does visualize a sort of picture being sent across the galaxy—but a picture that is based on mathematical concepts and one that cannot be distorted by the differences mentioned by Braude. We owe this concept primarily to Frank Drake. Just as Drake set the standard for most of today's listening programs with Project Ozma, so he set the standard for most of today's concepts of interstellar messages.

The first scientific meeting on intelligent extraterrestrial life in the United States was held at Green Bank in 1961. Drake was one of the participants. At the end of the meeting, he sent each of the other participants a message consisting of 551 zeroes and ones, in what seemed to be random order. His challenge to them: decode this message.

To the participants in the conference, the first step in decoding was simple. There were 551 characters in the message, and 551 can be divided by only two numbers: 19 and 29. Therefore the characters had to be arranged either in 19 groups of 29 characters or 29 groups of 19 characters. The second step was also simple: a zero was taken as a white background square and a one was taken to be a black message unit square (reversing the color scheme does not change the basic idea). When the squares were arranged in the proper array, figures and some regularly ordered sequences are apparent. The third step in the decoding process is to determine the meaning of the figures and the sequences. Some of the participants did, some didn't. Still later, Carl Sagan, Drake's partner, asked some scientists to decode a message that was similar to the one devised by Drake in every respect but one: a single zero was missing. No one could make sense of the resulting jumble.

Leaving out a single pulse eliminates one of the most important clues to interpreting the message: the fact that the number of characters in the message is divisible by only two numbers. But there are other pitfalls, as can be seen by an effort to decode a very similar message, one which has a place (however small) in the history of communication with other planets. This latter message contains the same sort of information that was transmitted in Drake's original 551-character message. It is unusual because of the recipients for whom it was intended.

DRAKE'S MESSAGE: "AN ANTI-PUZZLE"

Always willing to pioneer, Drake has the distinction of being the first person on earth to transmit a coded message toward the cosmos. The message was transmitted at 1:30 P.M. on Saturday, November 16, 1974. The occasion was the rededication of the Arecibo radio telescope, whose 1,000-foot dish had been resurfaced to increase its sensitivity. To send the message, the radio telescope was used as a transmitter. In theory, it can communicate with an antenna of equal size anywhere in our galaxy. However, the message was sent toward another galaxy, a globular cluster known as M-13 which contains some 300,000 stars and is 25,000 light-years from earth. Three minutes were needed to transmit a total of 1,679 pulses consisting of two alternating whines in the radio signal. Drake called the message "an anti-puzzle," adding that it is "a code designed to be easily broken."

The message is shown in Figure 10, a pictogram which was produced by arranging the 1,679 pulses in an array of 23 by 73 characters—they are the only two numbers by which 1,679 is divisible—and coloring the "on" pulses black and the "off" pulses white. It is an anti-puzzle that needs some explaining for those lacking the proper training.

The language of the message is that of binary arithmetic, in which there are only two numerals: 0 and 1. The top row of the message tells how to count from one to 10 in binary numbers. Just below the binary arithmetic lesson is a group that can be recognized as five numbers—from right to left, the numbers 1,6,7,8, and 15. These are the atomic numbers of hydrogen, carbon, nitrogen, oxygen, and phosphorus, all of which are essential to life on earth.

Assuming that an alien civilization now understand both the language of the message and the essentials of chemistry, the anti-puzzle gets seriously into biochemistry. There are two rows which picture the components of nucleic acids—a phosphate group, a molecule of deoxyribose, and the formulas for thymine, adenine, guanine, and cytosine. Put them all together and they form DNA. An explanation issued by the Arecibo observatory before the message was sent said that "knowledgeable organic chemists anywhere should be able through rather simple logic to arrive at a unique solution for the molecular structures being described here." This assumes, of course, that DNA is the molecule that contains the genetic information of the species that receives the message, just as it is the genetic molecule of life on earth. The curlicues in the center of the message represent the helix in which the DNA molecule occurs, and the block of squares in the middle of the helix gives the binary number for four billion, the number of nucleotide pairs in human DNA.

Now the message abandons pure arithmetic to show a crude picture of a human being, which is flanked on the right by a line running from its head to its feet and the binary number 14. This says that the human is 14 units tall; the species decoding the message is expected to know that the unit is the wavelength of the transmission, 12.6 centimeters, which would make the figure 5 feet 9½ inches tall. On the left of the figure is another binary designation of the number four billion, the number of people on earth at the time the message was sent.

Next comes a rough sketch of the solar system, with the sun to the right and the nine planets strung out toward the left. There is some attempt to indicate the relative sizes of the planets, and the figure for the third planet is displaced slightly toward the human figure to indicate that it is the home planet of the species that sent the message. At the bottom there is the image of a radio telescope, and an indication of its size—2,430 wavelengths across, or 1,004 feet.

"This information tells indirectly, when taken with the strength of our signal, a great deal about the level of our technology," the observatory's explanation said. "If the message is transmitted repeatedly, a desired impression is made that the message emerges from the telescope."

A great deal was made of the Arecibo message, both before and after it was sent. Drake and his colleagues tried out various forms of the message on a number of scientists to see how well they could decode it. Most of them understood the greater part of the message; whether an intelligent species in another galaxy would have any success at all cannot be said.

Those in favor of the message described it as historic. But there was grumbling because the international SETI community had not been consulted before a message was sent. After all, there had been an informal agreement at the meeting in Armenia that no one nation should start communicating with other civilizations on its own. A Soviet report in 1976 expressed the hope "that there will be further international discussions on the subject before such a program of transmissions is launched."

Actually, Drake indicated in an interview not long after the telescope rededication that the message was regarded more as a symbol than as a serious effort at interstellar communication. The frequency chosen for the signal was not any one of the recommended lines that have been discussed so frequently. While the observatory explanation spoke of a message being "transmitted repeatedly," it was in fact sent only once. And it was sent not toward any nearby sunlike star but toward a galaxy at which the antenna happened to be aimed at the time of the rededication ceremonies. Since about 50,000 years would be needed for the message to be received, understood, and answered by an intelligent civilization in M-13, the breach of earth's security by the Arecibo adventure is not calamitous. If the recipients come toward us with evil in mind, earth has a lot of time to prepare for them.

MESSAGE TO JUPITER

We will have even more time to prepare for an earlier message that was launched toward the cosmos, the plaque that was placed on the

Pioneer 10 and 11 spacecraft. Pioneer 10 was launched on March 3, 1972, for its mission to Jupiter. Before the launch, late in 1971, a science writer named Eric Burgess pointed out to Carl Sagan that Pioneer 10 would eventually leave the solar system, and so could carry a message from earth to outer space. Sagan got in touch with the appropriate authorities and was told, to his surprise, that no one objected to the idea of putting a message aboard the spacecraft. Within three weeks, the plan for what Sagan later called "a message in a space-age bottle" was proposed, approved, and carried out.

The result was the plaque that is shown in Figure 11. Its message was the result of a quick conference between Sagan and Drake; the human figures were drawn by Sagan's wife, Linda Salzman Sagan. The figures were etched on six-by-nine-inch plaques that were attached to the antenna support struts of both Pioneer 10 and Pioneer 11, which was launched in 1973.

At the top of the plaque is a representation of two hydrogen atoms, drawn so as to indicate to a scientist that they are emitting 21-centimeter radiation. The binary number 1 is beneath the atoms. At the left center is a pattern of radiating lines, each line with a long binary number. As Sagan explains it, if the numbers are interpreted as meaning distance, the diagram makes no sense, since the distances (in terms of the binary number times the 21-centimeter marker) are too small by interstellar standards. But if the numbers are interpreted as time intervals, they are supposed to make sense to an astronomer: the lines represent pulsars, and the numbers represent the time between emissions of radio-wave bursts by each pulsar. The length of the line represents the distance from each pulsar to earth, which is at the point where all fourteen lines converge.

By using the pulsar segment of the diagram, an extraterrestrial astronomer will be able to locate the source of the message. In addition, the astronomer would also be able to determine when the message was launched. The period of each pulsar is indicated with great precision, as of the time of launch. But pulsars run down slowly, so that the interval between bursts grows longer as time goes on. By comparing the current interval with that indicated on the dial, an extraterrestrial astronomer could hit on 1972 as the year of the spacecraft's launch.

The rest of the message consists of pictures: at the bottom, a picture of the solar system, with binary numbers indicating the scale, and a schematic representation of the spacecraft leaving the earth. At the right are a nude couple, with an outline of the spacecraft behind them and binary numbers to the right indicating their size.

Naturally, the nudes got most of the attention. The *New York Daily News*'s headline on the story said: "Nudes and Maps Tell About Earth to Other Worlds." There were complaints from some people that (a) putting a nude on a spacecraft is obscene; (b) the woman is pictured as passive while the man is active, which is not equality; (c) the figures look too Caucasian; and (d) that no extraterrestrial species could make any sense out of the drawing. An unauthorized version of the plaque that circulated during the Pioneer 10 encounter with Jupiter showed the man raising more than his right hand in friendship.

Citizens wrote letters and newspapers wrote editorials about the significance of the plaque. The great room that exists for determining its significance is indicated by the fact that Sagan later referred to the Pioneer message as "mankind's first serious attempt to communicate with extraterrestrial civilizations," "a message to back here" and "good fun." The last idea deserves some defense. If Pioneer 10 were traveling to the nearest star, it would arrive in 80,000 years. In fact, it is going nowhere in particular. By chance, it will travel toward a part of space where there are no important objects. Since the plaque is made of gold-anodized aluminum and the erosion rate in outer space is low, the plaque should last for many millions of years. Thus, if there are intelligent species out there and if they are roving the void of interstellar space, they might eventually find the plaque. The chances of that happening are infinitesimally small. Still, the thought of a virtually indestructible message from humanity sailing off into the cosmos intrigues many people.

The concept is so intriguing that the National Aeronautics and Space Administration did it again when two Voyager spacecraft were launched on a Jupiter-Saturn mission in 1977. This time the message was far more elaborate than the one carried by the Pioneer spacecraft, although whether it represents an improvement can be debated. This was a space message that was decided by committee.

PHONOGRAPH RECORDS IN SPACE

Instead of a plaque, the message is carried by a phonograph record, a twelve-inch copper disc made to be played at 16⅔ revolutions per minute. For the benefit of any extraterrestrials who come across the record, each of the Voyager spacecraft also carries ceramic phonograph cartridges and a needle; the whole package is in an aluminum container upon whose lid are pictures showing the speed at which the record should be played and the way that the cartridge should be used.

The record itself begins with a long train of signals that can be interpreted to form pictures. To start with, the picture sequence places the earth in space, first showing the solar system and then moving in on our home planet. After this lesson in astronomy comes a lesson in terrestrial biology. There are pictures of DNA and of living cells, and then the sequence moves on to human anatomy.

However, to the great amusement of NASA-watchers, the space agency drew the line at showing human reproduction. Originally, the sequence was to have included some clinical diagrams showing how babies are made and born. Remembering the reaction to the nudes on the Pioneer plaque, the committee that designed the Voyager message kept it as conservative as possible. The sequence of human reproduction included only a man and a pregnant woman, with a diagram showing the location of the fetus. Even that was too much for the space agency.

Too smutty, said NASA, which eliminated the pregnancy sequence entirely from the plaque. A reporter who called Herbert Rowe, NASA's associate administrator for external affairs, was told that the pregnancy picture "was not appropriate for inclusion." NASA also drew the line at sketches showing human sex organs. Voyager went into space with a sexless human species aboard. It did, however, include the names of all the members of the House and Senate committees in charge of space science.

At any rate, any being who finds the record, plays it, and deciphers the pictures will get a full look at the earth and its inhabitants: a diagram of the earth's internal structure and another of continental drift; pictures of the Snake River and the Grand Tetons in the American West; the majestic Monument Valley in which John Ford has filmed many of his classic Westerns; pictures of trees, insects, animals, and birds; Jane Goodall observing chimpanzees; a selection of the world's different peoples, from an old Turkish man to a Balinese dancer; the Great Wall of China and the Sydney Opera House; a sunset with birds and the Arecibo radio telescope. And more.

The record also contains greetings in a variety of human tongues, fifty-five in all, from Akkadian to Wu, a Shanghai dialect. Akkadian, which heads the alphabetic list, is also the oldest language on the record. It was spoken by the ancient Sumerians and is probably the oldest known language. Then come greetings from Kurt Waldheim, Secretary General of the United Nations, and from President Jimmy Carter. After that is a sequence of the "sounds of earth," starting with the "greeting call" of a whale and including the eruption of a volcano, the sounds of rain and surf, and a train whistle, and ending with a kiss, the sound of medical instruments, and recorded pulsar sounds.

The record ends with music. In his book, *The Lives of a Cell,* Dr. Lewis Thomas says that if he were to choose a messsage to space, "I would vote for Bach, all of Bach, streamed out into space, over and over again." In fact, the record does begin with Bach—the first movement of the Second Brandenburg Concerto. It is followed by Javanese gamelan music, percussion music from Senegal, a Mexican mariachi band, and so on to the concluding selection, the great Cavatina from Beethoven's Thirteenth String Quartet.

LAUNCHING BOTTLES INTO THE COSMIC OCEAN

You could say, as Carl Sagan did, that "the launching of this bottle into the cosmic ocean says something very hopeful about life on this planet." Or you could say that the message doesn't mean very much in practical terms. Voyager 1 will not leave the solar system until

1987, when it passes the orbit of Pluto. Voyager 2 will be about two years behind. In about 40,000 years, both spacecraft will pass within one to two light-years of a minor star. There will be other near approaches to stars, 147,000 years and 525,000 years after launch. If technical societies survive on earth and continue to grow in the way that they have in the past, it is probable that high-velocity space probes, perhaps with human crews aboard, will be roaming through the galaxy by that time, or that a full-time interstellar beacon will be transmitted from powerful terrestrial antennas. If today's technological society comes to a sticky end, sending a message by spacecraft will not accomplish anything. Either way, logic says that we should wait for a while.

However, logic doesn't seem to have much to do with it. The meaning of these messages is symbolic. They keep the SETI spirit alive until some more meaningful steps can be taken.

In reality, the clearest messages that earth has sent into space have been unwitting. We have sent a variety of them, all easily decodable by any intelligent species that happens to intercept them. They are the normal radio, television, and radar signals emitted by transmitters who have no idea at all of signaling to another civilization.

Very faint traces of the first radio transmissions made on earth now are about 80 light-years from us. These are not very important as interstellar signals because most radio waves bounce back from the ionosphere, so that the radio signals sent into space are barely detectable. But signals in the microwave region go out into space almost unhindered. Television and radar use frequencies in this region of the spectrum; tall antennas are needed for television stations because the signal is not reflected back to earth as the transmissions of radio stations are. Several people in SETI have pointed out that the earth is now surrounded by a growing bubble of electromagnetic radiation that carries the full selection of early television broadcasts. In addition, there are radar signals, the most powerful of which come from the huge systems built for military purposes. These are our first messengers to space.

In 1978 astronomer W. T. Sullivan of the University of Washington, assisted by two undergraduates, made what appeared to be the first detailed study of such signals from earth. Sullivan concluded that the most powerful signals which we are emitting come from BMEWS—ballistic missile early warning system—radar sets, which can be detected at a distance of 250 light-years by a society with our present technology. However, Sullivan said that BMEWS transmissions are not likely to give much information to a distant observer, for two reasons: there are only a few BMEWS transmitters, and their frequencies are shifted constantly for security reasons. At best, Sullivan said, BMEWS radar systems "may well act as 'acquisition signals' to the outside observer, announcing our presence, but yielding only a minimum amount of further information."

Television stations are a much better source of information about the earth, Sullivan said. There are about 15,000 television transmitters on earth, more than 2,100 of them with effective power of 50 kilowatts or more. Nearly half the transmitters are in North America, about 30 percent are in Western Europe, 6 percent are in Eastern Europe, and the rest are elsewhere. An observer who kept careful watch on the way in which the signals from these stations appeared and vanished with the turning of the earth could get a great deal of information, Sullivan said:

> After several years of careful monitoring of the intensity and frequency variation of several hundred stations, the observer could deduce (i) the complete orbit of the earth; (ii) the existence of station broadcast schedules influenced by the sun; (iii) the presence of an ionosphere and perhaps even a troposphere; (iv) the size, rotation rate and axis of rotation of the earth; (v) a complete map of the stations; (vi) the mass and distance to the moon; (vii) and the size of the radiating antennas; and (viii) various cultural inferences concerning our civilization.

However, this wealth of information is not widely available to listeners in the galaxy. Sullivan calculated that the earth's television signals can be detected by listeners at our level of technology only within a radius of 25 light-years. There are about 300 stars in that volume of space—a large number in the abstract, but a small number by the standards of the SETI formula.

These signals are not negligible. In some radio frequencies, the earth radiates more energy than the sun. However, the consensus in

SETI right now is that these signals are not likely to be detected by another civilization. Even if another civilization is looking for signals, "leaked" radiation transmissions, such as those from TV and radar, are not very easy to detect. The carrier frequency of a television station is only one-tenth of a hertz wide, and it carries no information. To detect the program content of a television signal, a listener needs an antenna that is 1,000 times more sensitive than the one needed to pick up the carrier frequency. In addition, the 25 light-year radius in which these signals are detectable represents a negligible distance in a discipline that deals in hundreds or thousands of light-years. Finally, the earth's transmissions of such signals may lessen in the years ahead because more signals are being sent by cable, as in home cable television. A television signal sent over a coaxial cable does not go into space, and an increasing percentage of our signals are transmitted by cable. Today's microwave towers, which spill radiation into space, may also be replaced by cables made of clear glass, which transmit light waves and do not radiate into space. Telephone companies are experimenting with laser-wave transmissions because the shorter wavelengths of visible light can carry much more information than microwaves. The use of lasers will also reduce unintentional transmissions into space.

Right now, a properly turned viewer on a nearby planet would see the earth glowing brightly in a few radio frequencies. In several decades, the glow might wink out as corporations and governments move to more efficient modes of transmission. If it happens on earth, we can assume that the same thing has happened on more advanced planets. Therefore we cannot count on making accidental contact with another civilization that has picked up or leaked radiation. We cannot even assume that another civilization is searching for such leaked radiation because that civilization's scientists probably will have concluded that the search would not be successful. If another civilization has its antennas pointed toward earth, it is almost certainly trying to detect an interstellar beacon.

Catch 22: Earth cannot send such a beacon at this stage of our history, largely for practical reasons. The Arecibo radio telescope could theoretically send a signal to an antenna of similar size anywhere in the galaxy, but no radio astronomer wants the earth's largest radio telescope dedicated to endless transmissions to an unknown receiver. Even a smaller antenna, such as the 300-foot dish at Green Bank, is too valuable to use for sending interstellar signals.

In theory, we could build a 300-foot antenna and use it solely for sending signals to other stars where there might be intelligent civilizations. The study made for Project Cyclops calculated that a 300-foot antenna would be capable of picking up a signal from another 300-foot antenna 80 light-years away. In theory, the matter would be simple: just pick out all the sunlike stars within 80 light-years, encode the message, and beam the message at each of the stars for a few days at a time. The antenna could either have a twin that would survey the same starts for return transmissions, or it could spend half its time listening and half its time sending.

That's theory. In practice, sending an interstellar signal is not in fashion in SETI right now. One reason is the international complications that might ensue if only one nation began transmitting. A recent Soviet statement said, "Soviet scientists have no plans at present to send out signals addressed to extraterrestrial civilizations. They share Sir Martin Ryle's hope that there will be further international discussions on the subject before such a program of transmissions is launched."

THE FEAR OF CONTACT

Sir Martin Ryle won the Nobel Prize in physics in 1974 for his role in the discovery of pulsars and for other contributions to astronomy. In 1976 he wrote both to Frank Drake and to the International Astronomical Union proposing an international agreement not to send interstellar messages from earth. His motive was fear—fear that an advanced society would follow the beacon to earth with plunder in mind. Perhaps the civilization that received the signal would want to colonize the earth, or perhaps it would want earth's mineral resources, Ryle believes.

Most members of the SETI community do not share that concern. It is generally believed that any civilization which can travel across the vast distances of space will have no need for any of earth's resources, including its inhabitants. As the recent history of our planet shows, slavery makes economic sense only for societies in a low state of technological development. As for resources, the earth has nothing that an advanced civilization needs. Already there is talk on earth of mining the minerals of the asteroids, the flock of planetoids that swarm between Mars and Jupiter. Freeman Dyson has taken the idea a large step forward. He foresees the time when not only the asteroids but also the comets are pressed into the service of humanity. Ventures in astro-engineering can make comets habitable, Dyson has written, and the time will come when humans ride comets through outer space. With such an abundance of space resources available, what species would want to invade an inhabited planet?

The fear of contact is real. But it is not fear that has put interstellar transmission out of style in earth's SETI community today. The real reason is inefficiency. Robert E. Edelson, SETI project manager at the Jet Propulsion Laboratory, explained it this way:

> Where would we aim the signal? Suppose we chose the nearest sunlike star, about eight light-years away. If there just happened to be intelligent life around that star and if they just happened to be listening to the right frequency at the right time with their antenna pointed in the right direction, they would detect us eight years from now, and if they immediately replied, then we would receive their signal eight years after that. Now, that's an extremely unlikely scenario, and yet it would take sixteen years to receive a response.

Or as John Billingham, a SETI leader at the Ames Research Center puts it, if the civilization is 800 light-years away, it would take 1,600 years to get an answer. Those numbers indicate that sending signals doesn't make sense, at least in the present state of affairs. The idea of sending signals is far from dead, but it is dormant. For the time being, the big push in SETI is to get some major listening programs started. To listen on any large scale, money is needed—government money. It is the need for money that is the major shaping force in SETI today.

CHAPTER ELEVEN

STRATEGIES

Some time in the early 1970s, the field that had called itself CETI began calling itself SETI: Communication with Extraterrestrial Intelligence became the Search for Extraterrestrial Intelligence. The explanation for the change in the title was that "communication" implied two-way messages, and that the stress in SETI was on detecting signals, not sending them. There is nothing untrue in that explanation. But there was another, more subtle reason in the background. The change in acronym also marked a change in the point of view of most people in the field: a ripening, a growing maturity, a reining-in of some of the wilder and more expensive ideas. It was no coincidence that the change occurred in the years when CETI (or SETI) changed from a purely theoretical exercise to a discipline that had gained almost complete scientific acceptance and had begun to compete with other disciplines for a limited supply of money.

PROJECT CYCLOPS

Perhaps the best way to illustrate the change is to describe two proposed plans for the detection of interstellar signals that were made just a few years apart. The first, Project Cyclops, was published in

1971, when the field was still called CETI. The second study became public in what was called a prepublication form in 1977, and is titled simply *The Search for Extraterrestrial Intelligence*. Both plans were formulated after a series of workshops on various aspects of interstellar communication. Many scientists participated in both studies, and many of the statements in the 1977 report were taken bodily from the Project Cyclops publication. But there is almost no comparison between the programs recommended in the two plans.

In scope, duration, and expense, Project Cyclops ranks with Project Apollo, which put men on the moon at a cost in excess of $20 billion. A couple of sentences from the report gives an idea of its scale:

> From the air, the final Cyclops system would be seen as a large central headquarters building surrounded by an "orchard" of antennas ten kilometers (6.25 miles) to ten miles in diameter and containing 1,000 to perhaps 2,500 antennas. . . .
>
> Cyclopolis, the community where the system staff and their families live, might be located several miles from the array, perhaps behind hills that provide shielding from radiated interference. Transportation to and fro could be by bus at appropriate hours. Alternatively, the central headquarters might be made large enough to provide the necessary housing, stores, schools and so on. There would be ample room between the antenna elements for playgrounds and recreation facilities . . . (See Figures 12,13.)

The 243-page report on Project Cyclops envisioned a long-term program to build the orchard of antennas. Each dish would be 100 meters—or 330 feet—in diameter. There would be at least 1,000 and perhaps as many as 2,500 antennas in the orchard, giving a total antenna surface that would be measured in square miles. Cyclops would start small, with only a few antennas around the central headquarters, but it would grow steadily, with more antennas being built every year. The cost was envisioned at about $600 million a year for ten to fifteen years, a total of anywhere from $6 billion to $10 billion. And since such an array would not be able to scan the entire sky, the report said that two of them might be built, one for the northern hemisphere and one for the southern hemisphere. Operating costs for a Cyclops system would be in the hundreds of millions of dollars a year.

For this, the report said, taxpayers would get a system that could scan the sky 200,000 times faster than Project Ozma. The report, written by Bernard Oliver, said that the comparison is made

> not to disparage Ozma, but to build faith in Cyclops. Ozma cost very little and was a laudable undertaking, but the power of the Cyclops search system is so enormously greater that we should completely discount the negative results of Ozma. The Tau-Cetacians or the Epsilon-Eridanians would have to have been irradiating us with an effective power of about 2×10^{13} watts (2,000 billion watts) to have caused a noticeable wiggle of the pens on Ozma's recorders; 500 kilowatts would be detected by Cyclops.

The report envisioned a four-phase operational scheme for Cyclops. In the first phase, during the planning stage, a "target list" of stars to be studied would be compiled by using a battery of light telescopes operating under automated control. The telescopes would divide the sky into segments, each telescope recording the visual spectrum of all the stars in its segment. Each star would automatically be classified by spectral group, so that only sunlike stars would be on the list to be scanned by Cyclops. From three to six light telescopes would be needed to prepare the target list, the report estimated.

Then comes the construction period, where about 100 antennas would be built annually for at least ten years. The search for extraterrestrial life would begin as soon as the first few antennas went into operation. "During the first years, the nearest stars would be searched for both leakage and beacons, and the techniques for doing this efficiently would be developed and tested," the report said. "Then, during the remaining construction years, the search would be carried further and further into space." Lest the pace be thought slow, remember that the report envisaged a scan of 15,000 stars a year, with an observation time of 2,000 seconds per star. At the end of ten years, when the system was full-fledged, the early stars probably would be restudied, using the full capability of Cyclops.

Now we are into phase three, "total search with the complete system." The first one or two scans looked at stars as distant as 500 or 700 light-years. Now the search would be extended out to 1,000 light-years. Cyclops might also be used as a beacon, sending out messages for a full year. In that case, the nearest thousand sunlike stars would be studied for possible responses.

Carefully stating that the four-phase scheme is a "worst-case" plan, based on the assumption that no signal would be detected, the report says that the fourth phase would be a pause for regrouping. The possibility that the basic thesis of SETI will have been disproved by work with Cyclops is discounted. Instead, the report says that "the decision to build a long-range beacon must be faced"—and, by implication, that the inevitable choice will be to build a beacon. Cyclops would then return to its search while the beacon would send endless messages to the stars; both would be fully automated. There is no indication of an end to the search or the transmissions, but rather a hope that "because other races have entered a Phase IV, this phase may never be reached."

Needless to say, Cyclops would be more than just a tool for SETI. It would also be a valuable instrument for astronomical studies. The report mentions several possible studies. Cyclops could serve as antenna for transmissions from spacecraft sent to explore the outer solar system; it could, the report said, allow spacecraft to transmit from distant Uranus with the same efficiency as today's transmissions from Mars. As a radar facility, Cyclops could map every major object in the solar system, including the moons of Jupiter and Saturn. As a radio telescope, Cyclops would improve observations by a minimum of tenfold over today's capabilities: many more pulsars could be detected; the diameters of distant stars could be measured; stellar motions could be studied with precision.

The importance of Cyclops as a facility for ordinary astronomy has been stressed by several people in the SETI community who are concerned with one major future problem in the field: boredom. If Cyclops were used for SETI alone, it might be entirely possible for a bright young astronomer to come right out of college into

Cyclops, work there for an entire career and leave, grey-headed, half a century later, with nothing but negative results to report. The prospect of having only acres of filing cabinets with detailed information on millions of stars that do *not* have civilizations to show for a lifetime of effort does not beguile the average astronomer. Therefore, it has often been assumed that a facility like Cyclops would either be devoted to ordinary astronomy for a fairly large fraction of its operating time or that most astronomers would serve only a portion of their careers at Cyclops. But as Cyclops was described in the report, its mission was clear: "The sole intent has been to optimize its performance as an initial detector of intelligent transmissions."

Cyclops was heady stuff. It made for marvelous newspaper headlines, and it came equipped with beautiful artist's conceptions of the completed orchard of radio telescopes. It was only if you looked very hard that you could see a few sentences that could be regarded as a disclaimer, a concession that the Cyclops program was really playacting. Those sentences came in the introduction:

> . . . Both the extremely preliminary nature of this study and the great uncertainty as to whether a system of this type will be built in the reasonably near future force us to emphasize that presently contemplated expenditures in radio astronomy, radio mapping and deep-space communication facilities should be evaluated without regard to the content of this report. It would be extremely unfortunate if funding for any worthwhile projects in this field were to be deferred or reduced because of a mistaken impression regarding the imminence of a Cyclops-like system . . .

So we could say that the Cyclops project is a form of window shopping, the wistful peek at a dream car that a buyer indulges in before going out to buy the sedan that fits his budget. From this point of view, the most valuable part of the Project Cyclops report was the sections that never got much wide public attention, the detailed calculations about the design of the antennas, the receiver system, the transmission and control arrangements, the signal processing. After the publication of the Cyclops report, the SETI community had a solid document full of information ready for anyone with questions about the search for extraterrestrial intelligence.

And perhaps it is only now that the Cyclops proposal looks overambitious. At the time—just two years after the success of Project Apollo—the habit of thinking big in space projects was still prevalent. The weight of the big names in the SETI community was behind the Project Cyclops study, and $600 million a year did not seem like so much in the context of the billions spent on Apollo.

LUNAR AND ORBITING SYSTEMS

Certainly the next report on SETI systems was no smaller in scale. This one came from the Stanford Research Institute, which is only a few miles away from Ames and which was paid by NASA to evaluate three systems: Project Cyclops, a Cyclops-type array of antennas on the far side of the moon, and an orbiting SETI system that would be built on earth and deployed in space. The reason for going to the moon or into orbit would be to escape the terrestrial radio signals that have caused so many false alarms in SETI programs so far. The lunar antenna system would be shielded from the earth's signals by the mass of the moon, while the orbiting system would have a large disc built between it and the earth to act as a shield.

None of these systems would be cheap. Going to the moon is the most expensive alternative, not only because of the high cost of launching the system into space—about $350 a pound according to the Stanford Research Institute estimate, with a total launch cost in the neighborhood of $25 billion—but also because of the need to maintain a large lunar colony to run the equipment.

Two kinds of lunar systems were considered. One would use an orchard of movable radio telescopes similar to those envisaged for Cyclops. The second would use Arecibo-type fixed antennas that would be built in lunar craters. Since the moon has a large number of craters of all sizes, the lunar SETI system could either use one enormous dish fitted into a large crater or several smaller antennas in lesser craters. The cost of building the antennas would be much lower on the moon than it would be on earth, because the moon's low gravity—one-sixth that of earth—eliminates the need for massive support structures. The study found that the cost of Arecibo-type lunar antennas was very much lower than a Cyclops-type moon system because of the lower construction costs. (See Figure 14.)

The orbiting system came in at an even lower cost. For a major SETI program, going into space is more efficient than going to the moon. In fact, the Stanford Research Institute study found that a SETI system in space could actually cost less over the long run than an earth-based Cyclops system if the number of advanced extraterrestrial civilizations is relatively small—fewer than about 10,000 civilizations in the galaxy.

The reasoning runs like this: If there are very many advanced civilizations in the galaxy, a signal from one of them could probably be detected in a relatively short time using only a small orbiting system; by "small" we mean an antenna whose surface is about half a square mile in area. But if advanced civilizations are uncommon, a long search with a much larger antenna array will be needed. The major cost of any orbiting antenna system, large or small, is the cost of research and development, the design of the system, and the construction of the very elaborate rockets and life-support equipment needed to build something big in space. In short, if you decide to build an orbiting SETI antenna, you might as well spend the little bit extra needed to build a big one, if you assume that the search for signals will go one for a long time. "For the very large systems that would be required . . . space systems are actually cheaper than earth-based systems," the report said. "Cheap" means $9 billion, in 1976 dollars.

The space system would have a "maypole" design, something like a giant umbrella. "It consists of a long central column hub, a rigid outer rim, and a system of cables [spokes] that tie the hub and rim together. A spherical mesh dish is suspended inside the 'wheel' to form the reflector." Think of it as an umbrella whose diameter would be about two miles. The different parts of the antenna would be manufactured on earth and put together in orbit in the form of a closed umbrella—the antenna surface folded down around the central column. When the antenna system was placed in an appropriate location, the umbrella would be opened, so that the antenna surface would be aimed in the desired direction, toward the stars.

The most likely spot in space for a SETI system is one of the

Lagrange points in the earth-moon system. A Lagrange point, named for the eighteenth-century French mathematician who first described the concept, is a point in space that defines an equilateral triangle (one with equal sides and angles), with the earth and the moon as the other two points. Lagrange developed the idea through observations of some asteroids that had an unusually stable relationship with Jupiter. A spacecraft, or an antenna system, placed in one of the earth-moon Lagrange points, would be equally stable relative to the earth.

Picture an antenna three miles in diameter floating at a Lagrange point. Smaller umbrellas of the same design would float above it, attached by booms, to gather signals from space and feed them to the antenna. To prevent interference from radio sources on earth, a shield would have to be placed between the antenna and earth. Without going into great detail, the Stanford Research Institute report said that the shield would have to be about six miles across, twice the diameter of the antenna itself. (See Figure 15.)

THE BUSINESS OF NASA

When an idealized project begins to assume tangible form, differences of opinion are likely to appear. So it is with SETI. Discussions showed that the SETI community was less enthusiastic about an orbiting system than NASA planners were. NASA's business is space projects, and an orbiting antenna is a major space project; something that could keep aerospace engineers working for years. The Space Shuttle, which is planned as the space workhorse of the 1980s, can easily move large quantities of material into earth orbit. But a new vehicle would be needed to move that material—maybe 8,000 tons of it—out of earth orbit to the Lagrange point to construct the space antenna. NASA would like nothing better than to build a "space tug" to do that job. The people who put men on the moon love the thought of space tugs and three-mile-diameter orbiting antennas.

SETI people are usually supported by NASA funds and often work for NASA, but they are not NASA people in the strictest sense of that phrase. SETI is primarily science—not entirely, as we shall see, but primarily. The major interests of most NASA people is hardware. Anyone who has been in the Vehicle Assembly Building at the Kennedy Space Center in Florida can understand the fascination of space in terms of hardware. The VAB is the world's biggest box—fifty-eight stories high. It was used to hold the world's biggest scale models, the equipment that sent men to the moon. Standing in the building and looking at the Saturn V rocket, which stood as high as the Statue of Liberty, one could understand the joy that comes from building a 1:1 scale model of the world's biggest rocket. You could also understand why Apollo 17, the last moon mission, was launched at night. The men who had built the machine wanted to see what it would look like at night. (It looked like a giant, cold arc lamp, yellowish-white.)

In the middle of the 1970s, signs of friction between the Ames Research Center and the Jet Propulsion Laboratory began to appear. Most SETI work had been done at Ames, although not all SETI people, by any means, were stationed at Ames (Cornell University, which has both Drake and Sagan, is a major SETI center in the East by virtue of their presence). JPL did some work on extraterrestrial intelligence, but not nearly as much as Ames. In the boom times of the 1960s, JPL was kept quite busy by other projects.

Then came the shrinkage of NASA funds in the 1970s. Competition for the limited amount of money that was available developed quickly. In that competition, JPL was a major winner: it was given complete supervision over the American program of planetary exploration. Ames, which had a number of significant planetary missions until then, was phased out of the planetary exploration program entirely.

So far, the division of responsibilities seemed to be clear: planetary exploration for JPL, SETI for Ames. But it did not remain clear. As the 1970s wore on, a growing interest in the field of SETI began to be evident at the Jet Propulsion Laboratory. In fact, when the time came for the SETI community to ask for official funding from the space agency, two separate proposals were presented to NASA: one from Ames and one from JPL.

Officially, the two programs were presented as complementary halves of a unified effort. However, at NASA headquarters, the Ames plan and the JPL plan were viewed as rivals for whatever money would be available for SETI in the 1979 fiscal year budget. From the beginning, it was obvious that either JPL or Ames would be funded, but not both.

The two plans presented quite different strategies in the search for extraterrestrial intelligence. The one thing they had in common was a de-emphasis of the think-big approach of Project Cyclops, for practical reasons. In the ebullient 1960s, it was easy to talk about projects that would cost tens of billions of dollars. In the cramped 1970s, that sort of talk was frightening and repellent to the administrators and legislators who drew up budgets and voted on appropriations. To a large degree, the SETI community found itself explaining away Cyclops as a mere adventure in speculation, a pull-back that was regarded as a necessity for getting any funding for a search for extraterrestrial signals.

"We have explained to everyone that Project Cyclops was a long-term study that was done some time ago," John Billingham of Ames said in a 1977 interview. "Since that time, everyone has gone back to square one, to fundamental questions. The outcome is clear. The way to start the program is at a much more modest level than Cyclops envisaged, to use existing radio telescopes and work at a very modest level before contemplating the business of getting into a dedicated system. [By that he meant a system totally dedicated to SETI.]

"Cyclops has sort of been pushed into the future. Cyclops is a big master plan that might be adopted someday. We're trying hard now to explain to people that nobody is proposing to build Cyclops. We are at a stage well before that."

SETI GROWS UP

You could put it another way: SETI was growing up. It was becoming more than a symbolic effort that a few enthusiasts would undertake only to prove that it could be done, more than a fantasy that could be outlined with no thought to time or money. The narrowing of objectives that was evident between the Cyclops report of 1971 and the SETI report of 1977 might have seemed disappointing to some, but it was really a sign that SETI was coming of age. The 1977 report was talking about a few million dollars now, not tens of billions of dollars in the misty future. It had to make a hard, businesslike case for an appropriation in a budget where money was not easy to come by.

What SETI lost from this approach was charm—the sort of charm that was lost when Apollo landed on the moon and ended all speculations about the composition of the lunar surface, or when Viking landed on Mars and substituted data for imagination. The great fascination that SETI holds for most people is a matter of charm. SETI has been Little Green Men and Unidentified Flying Objects and Chariots of the Gods and Dyson spheres and Kardashev civilizations and space probes and all the rest. Now it has come down to a matter of dollars and cents, of offering specific programs that give the maximum chance of finding another civilization at a reasonable cost. The loss of charm is regrettable but inevitable. UFOs retain charm because they are impossible, the stuff that dreams are made of. SETI is losing charm because it looks more and more like a real-life enterprise. The numbers that go into SETI calculations—the distances of the galaxy, the times of stellar evolution—indicate that contact will not be made in one quick, romantic burst of effort. SETI must settle down to the long grind.

The new 1977 SETI report reads like a prospectus. It is backed by two years of workshops covering every aspect of extraterrestrial intelligence, from star formation to electronic equipment. The participants in those workshops included not only all the big names in SETI—Morrison (who served as chairman), Bracewell, Drake, Sagan, Oliver, Billingham, and many others—but also a number of influential members of the American astronomical community. The report, edited by Morrison, Billingham, and John Wolfe of Ames, sets forth what is described as a "consensus," with four main conclusions:

(1) It is both timely and feasible to begin a serious search for extraterrestrial intelligence.
(2) A significant SETI program with substantial potential secondary benefits can be undertaken with only modest resources.
(3) Large systems of great capability can be built if needed.
(4) SETI is intrinsically an international endeavor in which the United States can take a lead.

Note the emphasis on practicality. SETI is a serious scientific subject. A program can be started with "modest resources"—a few million dollars, not many billions—and it will have "substantial potential secondary benefits." Yes, large systems like Cyclops can be built—but not now. Finally, if the United States takes the lead, other nations will help share the load: "We can and should expect growing cooperation with investigators from many countries, both those already displaying interest and activity, as the Soviet Union and Canada, and others whose interest would grow."

All of these statements are selling points, not only for a money-conscious Congress but also for the NASA administration and the astronomy community. Not all astronomers are keen on SETI; many of them regard it as a quixotic enterprise that will not add much to our knowledge. Shortly after the report began to get publicity, Jesse Greenstein of the California Institute of Technology, one of the most respected figures in American astronomy and a member of the committee that prepared the SETI report, was quoted as saying, "I'm not entirely sure I either like or approve of the enterprise."

Greenstein's doubt was based on scientific questions, but money is also an issue. Astronomy has largely become a government funded endeavor, and government money for science has tightened in the 1970s. Money that goes to SETI comes out of someone else's budget, directly or indirectly. Hence the stress in the 1977 report on the usefulness of the proposed SETI program for all astronomers: "As soon as a dedicated SETI facility achieves either a sensitivity or spectral coverage not found in present radio or radar astronomy instruments, it becomes a uniquely useful tool for research in these areas. An almost continuously increasing spectrum of applications exists as the SETI facility is expanded in scope." And hence the emphasis on low initial cost:

> A large, expensive system is not now needed for SETI. If we but equip existing radio telescopes with low-cost state-of-the-art receiving and data-processing devices, we will have both the sensitivity to explore the vicinity of nearby stars for transmitters similar to earth's, and to explore the entire galaxy for more powerful signals. . . . It will be timely to consider

whether to proceed with a larger-scale program after this earlier effort has shown us more accurately what might be involved.

Both the JPL and the Ames proposals fit those general guidelines, but the differences between them were basic. As Robert Edelson of JPL explained it, "They [Ames] want to look at stars that are likely candidates for having life. If you do that, you can get very good sensitivity. The approach we are taking is to look at the entire sky to try to detect weak signals. In our state of knowledge, we don't feel that there is any particular preference for selecting stars to be scanned." In other words, the old hands at Ames proposed to push ahead with the same strategy that had begun with Project Ozma: looking at specific stars and studying very narrow bandwidths. By contrast, the relative newcomers at JPL had a strategy resembling the Soviets': looking at the whole sky over relatively broad bandwidths to detect extremely powerful signals.

Both the Jet Propulsion Laboratory and the Ames Research Center said they would use much of the NASA money to build an innovative piece of electronic equipment called a signal processor, which would make the search for an intelligent signal more efficient than ever before. The signal processor would do its work after the antennas collected the radio waves. The Jet Propulsion Laboratory proposed to use rather small antennas: an 85-foot dish and a 30-foot dish in the desert at Goldstone, California, both part of NASA's Deep Space Network, and a 14-foot horn reflector. (See Figure 16.) JPL proposed to look methodically at a large portion of the sky in all the frequencies from 1,000 MHz to 25,000 MHz, taking slices of 300 MHz at a time. (For example, it might look first at the range from 1,000 to 1,300 MHz, then at the range from 1,300 MHz to 1,600 MHz, and so on.)

The signals from the antenna would go to preamplifiers that would separate signal from noise. Advances in electronic equipment will allow JPL to build preamplifiers better than any used before in SETI, yielding a cleaner signal to analyze.

After the preamplifiers comes the signal processor. It is designed as one of the wonders made possible by the recent explosion in microelectronics. Electronics engineers have been able to squeeze more and more components on a semiconductor chip. A single chip that can fit on a fingernail now has the power of a computer that would have filled a room in the 1960s. These chips are cheap; computer central processing units by the barrelful now cost about $20 a chip. (One hesitates to quote a price because any quote is too high by the time it gets into print.) Using such chips, JPL can put the 300-MHz-wide signal that emerges from the preamplifier through a processor that will look at 100 million different frequencies simultaneously for the faint, coherent signal that could be a message from another civilization.

The JPL scheme would look at about 80 percent of the sky—all that can be seen from the Goldstone site—one patch at a time. Edelson calculates that the search will take five years and will take in perhaps 100 million sunlike stars. Most of the data will be thrown away because, as Edelson says, "We could fill the Library of Congress with books of this output in just a fraction of a day." Anything that looks interesting will be retained on magnetic tape—and "interesting" means signals of astronomical importance as well as suspected messages from other civilizations. Like the ongoing Ohio State scan, the JPL scan will single out a number of objects in space that are worthy of further study by astronomers. As Edelson explained, "Even negative results will be significant."

The Ames strategy is more conventional. (It is remarkable to think that SETI is now old enough so that a strategy can be described as "conventional.") It would use very large antennas—the 1,000-foot dish at Arecibo, the 175-foot antenna at Ohio State, a similarly large but as yet unchosen dish in the Southern Hemisphere, either in Chile or in Australia. It would look at carefully selected stars, all of the 500 sunlike stars within 80 light-years of the earth. It would look at the well-established frequencies in the water hole. And it would look at slivers of the water hole only one MHz wide. Its signal processor, like JPL's, would analyze one million different channels in the signal. But while JPL would be looking at channels 300 MHz wide, Ames would examine individual signals only one hertz wide—roughly the width of the carrier band used by a television station.

In an initial search, Ames would look at each star for ten or fifteen minutes. Since its search would use only a small part of the operating time of the radio telescopes, the entire search would take between three and four years. If the first search was negative, Ames had plans for going back to look at different frequencies, perhaps using a more advanced signal processor that could search one billion channels at a time.

When the federal budget appeared in January 1977, JPL was the winner. The budget proposed $2.1 million as a start for a seven-year, $14-million broad-band search of the sky. Ames got nothing but a statement that its plan would get serious consideration in the future.

The people at Ames were unhappy and resentful. "Ames has been working on this for nine or ten years," one of them said. "We have done a lot of things in this field. JPL is just a latecomer."

More than that, Ames viewed the JPL plan as having very little value for SETI. People at Ames could pull out charts to demonstrate that the JPL program could detect only overwhelmingly strong transmissions from relatively nearby civilizations. From the Ames point of view, NASA supported the JPL plan not because of its value to SETI, but because the work done by JPL would help refine the operations of the Deep Space Network.

Mark A. Stull of the SETI program office at Ames put it this way: The Ames listening program could pick up signals from the most powerful transmitters now operating on earth—the BMEWS military radars—at a distance of ten light-years. "We could clearly detect the earth at ten light-years," Stull said. "And if earth had a lot of transmitters just slightly more powerful than the majority now operating on earth, we could see it out to twenty-five, fifty, a hundred light-years distance." By contrast, the JPL listening program is fifty times less sensitive than Ames's, Stull said. As Charles A. Seeger of Ames put it: "They [another civilization] would either have to have incredibly powerful transmitters, using planetary-size energy sources, or they'd have to know where we are to aim at us" in order to be detected by JPL's antennas.

SETI: SOVIET STYLE

Nevertheless, the JPL proposal did win the first round. That victory is less surprising if we look at the only other national SETI program now going, that of the Soviet Union. The Soviets seem to be following the same approach as the United States has chosen—they have started with a low-sensitivity listening program but have plans for an Ames-like system in the future.

In 1975 the journal *Soviet Astronomy* printed a report from the USSR Academy of Sciences Board of the Scientific Council on the Radio Astronomy Problem Area which included the statement, "Efforts to detect extraterrestrial civilizations should proceed smoothly and systematically, and should extend over a prolonged period of time. . . . It would be a great mistake to build a program in contemplation of rapid and easy success."

The Soviet report outlined two instrumentation projects. The first, running from 1975 to 1985, consists of the existing ground-based system for continuous monitoring of the entire sky, two satellites that would also be used for continuous, wide monitoring and a scan of nearby galaxies using relatively inefficient antennas. The Soviet strategy thus clearly resembles the JPL plan, except that the Soviets favor an even broader-gauge approach. If the earth is being bathed in powerful signals from an extremely advanced civilization, the Soviets could be the first to detect it. The influence of Kardashev can be seen in this approach. The Soviets are serious about the possibility that Type III civilizations exist. Their plan says that "The possible discovery of probes sent from extraterrestrial civilizations and now located in the solar system or even in orbit around the earth warrants particular attention." It also includes a proposal to search for "thermal radiation inherently emanating from large-scale works of engineering that may have been constructed in circumstellar space," a description that could fit either an advanced Kardashev civilization or a Dyson sphere. Such speculative ideas do not get much stress in the proposals of American astronomers, who want to avoid the accusation that they are dealing in science fiction, not science.

The second part of the Soviet plan, running from 1980 to 1990 (and thus overlapping with the first project), is slightly vague. But it does resemble some of the advanced American projects. It calls for:

(1) A satellite system continuously monitoring the entire sky and equipped with antennas of large effective area.

(2) A system of two widely spaced stations having large (effective area about one square kilometer) semirotatable antennas for synchronized reception, searches for signals from specific objects, and analysis of selected sources.

These instrumentation complexes could be used not only for CETI [the Soviets still use the old acronym] work but for a variety of important astrophysical problems.

In addition, additional parts of the program could be carried out with other radio telescopes in conjunction with the plans of radio-astronomy institutions (sky surveys, investigations of peculiar sources, and so on).

Even with its careful committee-written prose, there is more of the wild blue yonder in the Soviet report than in the latest American proposal. Part of that could be attributed to the Slavic temperament. Soviet science has always been rather open to what could be called borderline concepts: extrasensory perception, for example, has gotten more serious attention from the Soviet establishment than from ours. Robert Toth, a correspondent for the *Los Angeles Times,* was detained by the KGB for several days in 1977 for receiving documents from a Soviet ESP researcher. It is known that the Soviet military looked carefully at ESP for some time before concluding that it is worthless for military purposes.

In addition, the fact that Soviet budgetary decisions are made by small groups of individuals, rather than by the large groups of legislators, working in the glare of publicity, who approve the United States budget, means that an offbeat idea can get somewhere if only one or two key individuals are convinced of its worth. As the American SETI community waited nervously in late 1977 for the reaction first from NASA headquarters, where the agency's budget was being prepared; then from the White House, where the NASA budget would be pruned; and finally from Congress, where the proposed budget would be put through the congressional mill, its members might have yearned for older days, when science did not depend so completely on government. In times gone by, an astronomer who needed the equivalent of the $20 million that is being asked for the American SETI program would have gone to a sympathetic millionaire, not the government. Now such a proposal must meet the needs of consensus calls for a whole-sky, low-sensitivity study, not an effort to zero in on specific stars which could harbor civilizations that might be trying to get our attention.

And even that limited program is in trouble. The vision that exists within the SETI community is not shared by many members of Congress. In 1978, Senator William Proxmire announced that the SETI program would get one of his "Gold Fleece" awards, which Proxmire established to ridicule government-funded scientific programs which he regards as absurd. More important, both the Senate and the House subcommittees which considered the SETI request cut it drastically, from $2.1 million to $700,000. The cut amounted to an indefinite postponement of the proposed sky search. It was the most disappointing reception possible.

CHAPTER TWELVE

FUTURES

Maybe there is nothing out there.

If there is something, maybe we can never detect it.

Those two propositions are the dark side of the search for extraterrestrial intelligence. The pessimistic view that we are, in fact, alone in the galaxy is rarely discussed at length because people don't want to hear it. But the cause against SETI is made occasionally, and it must be considered.

One way to be pessimistic about life elsewhere in the universe is to attack the principle of mediocrity. If the earth is somehow unique, then the human race might either be the only intelligent species in the galaxy, or it might be one of a very few intelligent species that are so distant from one another that contact is effectively impossible.

The principle of mediocrity can be questioned at a number of points. For example, there is the assumption that in most planetary systems, at least one and probably more planets will be hospitable to water-based life. But in the solar system, we see that Venus cannot support life because of a runaway greenhouse effect and that Mars is at best a borderline habitat because it is too small to maintain an adequate atmosphere. As for earth, there are indications that a disturbingly small alteration in some of this planet's characteristics could make it quite unsuitable for life. For example, it has been estimated that displacement of the earth only six million miles closer to the sun would result in a runaway greenhouse effect. The earth would then receive so much solar energy that all its water would vaporize. Without liquid water, carbon dioxide would not combine chemically with surface rock, but would remain as a gas. Both water vapor and carbon dioxide gas absorb infrared radiation powerfully, so that earth would become the hell that Venus is.

THE DOUBTS

Then there is the SETI assumption that an intelligent species more or less like the human race will arise on most life-supporting planets. As we have seen, George Gaylord Simpson, one of the best paleontologists in the world, has expressed serious doubts about the inevitability of the evolution of intelligence. Some recent discoveries about the evolution of the hominids—the primates that evolved into *Homo sapiens*—indicate that our arrival on earth was the result of a particular combination of circumstances involving geography, climate, and animal evolution.

As the present picture runs, small primates had evolved from an ancestral mammalian stock in a dense forest that covered Europe, India, East Africa, and the Middle East. About 12 million years ago, for reasons that are not well understood, global temperatures began to drop and the thick forest was partly replaced by open grasslands. A few primates ventured out onto the open savannah, where evolutionary pressures led them to develop upright walking, dexterous hands, and all the other capabilities that brought about the evolution of the human race.

But suppose the climate had not changed just when it did. If the global cooling had occurred a few million years earlier, there might not have been primates ready to explore the new evolutionary opportunity. If the earth had not cooled at all, the forest-dwelling primates might have stayed in their secure ecological niche, without the need to develop toward human capabilities.

Another possibility that got a good deal of attention at one time centers on the role of the moon in terrestrial evolution. The earth-moon system is decidedly unusual in the solar system; no other planet has a satellite that is so large by comparison. The moon causes tides. It has been suggested that if there were no moon, the absence of tides would make it more difficult for marine animals to move out of the ocean to dry land. Therefore, if intelligence did evolve on a tideless planet, it would be the sort of intelligence postulated for the dolphins—that is, a nontechnological, water-dwelling intelligence. According to this train of reasoning, a moonless planet could not have an intelligent species with which the human race could communicate.

It is an interesting thought, but one which most scientists now reject for at least two reasons. Biologists are not at all certain that large tides are needed to help sea animals make the transition to dry-land living. And if they are needed, the tides caused by the gravitational pull of the sun could well be enough. The moon's gravity is responsible for only two-thirds of the earth's tides. If there were no moon, the earth would have tides one-third as high as the ones we have now. So the uniqueness of the earth-moon system does not mean that the evolution of an intelligent, technological species on this planet is a unique event.

Arguments against the principle of mediocrity are almost endless. But there is a single counterargument that is remarkably effective: an argument based on the vastness of the universe and the enormous number of stars like the sun. With so many stars and so many billions of years, the counterargument runs, it is highly improbable that the sun should be alone in having a planet that supports intelligent life. It is hard to believe that in all the roles of the cosmic dice, only one throw has been successful.

However, the most effective argument against the optimistic view of SETI is based on cosmological numbers. The pessimists say that the vastness of the universe is such that there is an almost vanishingly small chance of two intelligent species being close enough in both time and space to establish communications with one another.

As we have seen, "travelers" such as Bracewell, Kuiper, and Morris have used this argument to support their belief that space probes, not radio telescopes, are the most appropriate method for interstellar communication. The scenario that the pessimists write is always the same. It hinges on the improbability of two civilizations being at the same technological level at just the right time for both of them to be ready for interstellar contact. Time and space, the pessimists say, are both against SETI. D. R. Bates, a physicist at Queen's University in Belfast, Northern Ireland, has expressed his skepticism several times, as in a paper published in 1974 in *Nature:*

> . . . the number of communicative civilizations in the galaxy is at most 1,000, from which it follows that the likely distance to the nearest is at least several thousand light years. Perhaps most serious, the value of the . . . time before contact between a pair of dedicated radio telescopes is likely to be made becomes 10,000,000 years. . . . the prospect is unpromising.

But even if two civilizations that can communicate are relatively close to each other by galactic standards, we can still outline a pessimistic scenario that ends with no contact. Assume that the other advanced civilization is only 200 light-years from earth—an optimistic assumption in SETI. Since the earth has just arrived at the capability of extraterrestrial communication, we can assume that the other civilization is ahead of us. Let us say that it is at least 1,000 years ahead of us and that it has been working at interstellar contact for 500 years.

Civilization X, as we can call it, has a list of promising stars within 1,000 light-years of its home planet, and the sun is on that list. Five hundred years ago, Civilization X pointed its first signal-seeking SETI system at the sun and heard nothing. Four hundred years ago, the effort was repeated. Again nothing. Since then, Civilization X, whose prosperity grows as its technology advances, has surveyed the earth at ten-year intervals, each time for several days. All the results have been negative. They will remain negative for another 150 years or so, until earth's radar and television transmission bubble, traveling at the speed of light, reaches Civilization X.

Will it still be listening? That question is being debated right now in the planetary congress of Civilization X. Even though the planetary system of Civilization X is wealthy, some legislators are murmuring that the money which has gone into this centuries-old search for extraterrestrial intelligence could be better spent elsewhere—on an expansion of the asteroid-mining program, for example.

THE HOPES

But a pro-SETI legislator is rising to speak. More time is needed before the program is abandoned, the legislator says, pointing out that Civilization X began sending interstellar signals 150 years ago. An answer to those signals might arrive any day, the legislator says, and it is worth waiting for.

The hypothetical signal will arrive at earth in fifty years. Of course, it will fall on earth for a limited time only; each promising star is allotted a few days' beacon time. If the earth is listening, and if the message is received and understood, headlines will proclaim the first extraterrestrial contact. A return signal will be sent at once. It will arrive at Civilization X in 200 more years. Will Civilization X still be listening then? Highly unlikely, say pessimists, drawing on their knowledge of the short attention span of terrestrial legislators.

In an article in the magazine *Astronomy,* Gerritt L. Verschuur—who, if you remember, conducted the first search for extraterrestrial signals after Project Ozma I—concluded that SETI is a forlorn hope:

> It appears impossible for short-lived civilizations to contact each other, while long-lived civilizations are hundreds of light-years apart. Given man's present penchant for impatience, we can hardly claim to be communicative on the time scale required for interstellar discussions.
>
> The inevitable conclusion that evolution never stops leads to the assumption that L (the lifetime of civilizations in Drake's formula) is a small number. If this is the case, it follows that we are effectively alone in our little corner of the Milky Way. Civilizations light-years away might as well not exist. We shall never contact them; we will never find the needle in our interstellar haystack.
>
> We are alone in space!

THE FAITH OF SETI

Against this mass of pessimistic numbers, the SETI community musters a curious weapon: faith. To be sure, it is a scientific faith, bolstered by laboratory data, technical formulas, and the like, but it is faith nevertheless. A number of scientists have pointed out that belief in such borderline subjects as UFOs, extrasensory perception, and Kirlian photography can be described as an effort to substitute new scientific certainties for the old religious certainties that science has undermined. It has often been remarked that ufologists believe that the earth is being watched by beings from the heavens who have remarkable powers and who could save humanity from its present plight; the conclusion of the hit movie *Close Encounters of the Third Kind* certainly emphasized the religious tinge to the UFO belief. Erich von Däniken himself has said that religious uncertainty is a major reason for the success of his books, which describe the purported ancient astronauts as "gods." It must be noted that the same sort of faith is at the root of SETI.

If a SETI program on earth is to succeed in making contact with another civilization, that civilization will have to be very advanced, in ways that go far beyond its technology. It will have had to make a commitment to an extremely long-term effort at extraterrestrial communication. To fulfill that commitment, the civilization must have entered a prolonged period of affluence. Its leaders must have the vision to realize that success at interstellar contact requires unending vigilance and the devotion of a growing share of the civilization's resources. As for the technology of the civilization, it would have to be incredibly advanced by comparison with the earth's present capabilities to be able to transmit a beacon that the earth can receive—for it must be remembered that similar beacons would have to be sent toward many thousands of stars.

It might be an exaggeration, but not a great one, to say that the combination of technical mastery and wisdom attributed to such a civilization are almost godlike by the standards of today's squabbling, unhappy human race.

But even leaving any tinge of religion out of the discussion, the belief that such a civilization could exist is an act of faith in humanity. It is belief that a civilization like ours can break through all the problems that now threaten not only to bring our technological society to its knees but also to destroy the human race. The SETI community wants the human race to assume that such a positive breakthrough has taken place elsewhere in the galaxy. We are asked to spend first millions and then billions of dollars because of that assumption. It is stated in almost so many words by people in SETI that contact with such a civilization is almost certain to transform mankind.

"Underlying the quest for other intelligent life," Bernard Oliver wrote in the Project Cyclops report, "is the assumption that man is not at the peak of his evolutionary development, that in fact he may be very far from it, and that he can survive long enough to inherit a future as far beyond our comprehension as the present world would have been to Cro-Magnon man."

A VISION OF THE FUTURE

And remember Frank Drake's vision of the future, expressed in an issue of the magazine *Technology Review* that was dedicated to Philip Morrison: It is the year 1996, and after years of listening, an orchard of antennas in the Mojave Desert detected "a message in binary code, a message that will be received for a year before the scientists understand the format of the message and recognize it as the song of people who have been alive, every single one of them, for a billion years." The message, Drake's story goes on, carries the information that will make immortality possible for human beings.

> As never before, the people of earth now hold their destiny in their own hands. Should they adopt a life of immortality, thus causing a shattering transition in human life so great that its consequences are hardly predictable or imaginable? Or should they cling to death as well as their heritage of all that is good in human life?

Drake's story ends tantalizingly, with no answer to his questions: "It is the year 2030. The last humans remaining on earth . . ."

This is a vision literally of death and transfiguration, an unusual blend of science and something that can be described only as religion. We could draw a parallel by pointing out that the Christian religion envisions mankind's fulfilling its destiny in the heavens. So does SETI, occasionally in a manner as fervid as that of early Christianity. The young people who crowd into lecture halls to hear the latest negative results of searches for extraterrestrial intelligence are not interested in electronic hardware and budget requests nearly as much as they are in the vision of the future that SETI makes possible—a vision in which technology enables mankind, in the words of Carl Sagan, to "break the shackles of earth."

SPACE COLONY

A plan for getting mankind off earth and into space already exists. It was developed in the early 1970s by Gerard K. O'Neill, a professor of physics at Princeton University, and has been refined in a series of workshops and studies since then. O'Neill believes that a majority of the human race could be living in space colonies in just a few decades. It is a technological vision that has been enormously popular on college campuses, but decidedly less so among the decision makers at NASA. The very least that can be said about O'Neill's space-colony plan is that it is a dazzling proposal to leap over all of the problems facing mankind.

The plan would be possible only with regular flights of the Space Shuttle, a NASA spaceship that is scheduled to make its first orbital flight in 1981. The Shuttle will be a reusable launch vehicle, capable of putting up to 29 tons of payload into earth orbit with each mission at a much lower cost than has been possible until now. The Shuttle is a two-stage system. The first stage consists of rockets that put a winged vehicle into orbit; the rockets then parachute to earth to be used for future launches. The orbiter places its payload into orbit around the earth and flies back to make a landing at a specially prepared strip at the Kennedy Space Center in Florida.

For the first step in O'Neill's space-colony plan, several dozen Shuttle flights would put 2,075 tons of payload into orbit. Most of that payload would consist of Shuttle fuel tanks, which would be ground up in space to be used as an odd kind of fuel for a distinctively different kind of propulsion system called a mass driver.

Picture the mass driver as a slingshot. But instead of using rubber bands, the mass driver would have a series of buckets into which coils of superconducting wire are built. (A superconductor is a material in which an electric current, once started, goes on virtually forever. Superconductivity occurs only within a few degrees of absolute zero.) The buckets would be suspended above an aluminum track by the interaction of the current in the coils with the metal of the track. Electrical pulses can quickly accelerate the buckets to extremely high speeds. In space, the buckets would throw the ground-up material of the Shuttle's tanks backward and would thus achieve thrust. In a leisurely, spiraling journey, the 1,100 tons of ground-up tank would be used as fuel to place a 730-ton payload in orbit around the moon. The mass driver would then go back to earth orbit, ready to make the round trip again.

The earth-moon flights would establish a lunar mining colony. But the moon colony would not be an end in itself. It would be only a stepping stone to outer space, which O'Neill regards as mankind's destined habitation. The trouble with the moon, O'Neill says, is gravity: too much of it to make the shipment of moon minerals to earth profitable (rocket launches are expensive) but not enough of it to keep humans comfortable over prolonged periods. O'Neill believes that fewer than two dozen persons will be needed on the moon to run a permanent mining operation. Setting up that operation, his studies indicate, would require about 2,800 tons of Shuttle payload, or about 100 Shuttle flights. Of that total, slightly more than 1,000 tons would actually be landed on the lunar surface.

The mass driver itself would be included in the lunar mining operation—in fact, the mass driver would be the key element of that operation. The objective of the lunar mining camp is building a space colony. It would start by processing about 9,000 tons a year of lunar rock and would soon reach a capacity of over 600,000 tons a year. Moon rocks contain almost everything needed to build a space colony, O'Neill says. They are 20 percent silicon—which could be made into solar cells to turn sunlight into electricity—about 20 percent of a mix of metals; including aluminum, iron, and magnesium—which could be used to build the space-colony structure—and about 40 percent oxygen, which could be used to supply the space colony's atmosphere.

In the first stage of space colonization, the relatively small number of people in the space work crews would live in Shuttle tanks that would be converted to living quarters. The tanks in space would be connected by cables and would spin to provide artificial gravity for the inhabitants. The workers in space would be on the catching end of a cosmic baseball game. On the moon, material would be placed in the buckets of the mass driver and would be hurled off toward the temporary space camp. The materials would be caught and assembled into the first space colony suitable for large-scale habitation.

Several studies financed by NASA have produced a picture of that first space colony, which O'Neill calls Island One. It would be located at one of the earth-moon Lagrange points, L5. The first concept called for a torus, a doughnut-shaped space colony. More recent plans call for a sphere—called a "Bernal sphere" in honor of the British scientist—with a circumference just under one mile, and a population of 10,000. (See Figure 17.) Originally, O'Neill's plans called for Island One to be self-sufficient in food—plants could be grown in lunar soil sent up by the mass driver—but the latest studies say that the colony could be economically viable even if a large amount of food had to be brought up from earth.

Island One would make money by generating electricity. It could use large islands of solar cells to turn sunlight directly into electricity, or it could use mirrors to focus the sunlight, boiling water to drive turbines. Either way, the electricity would be sent back to earth as a powerful microwave beam. Within a relatively short time, O'Neill says, such space generating stations could supply an important fraction of the earth's electrical generating capacity. A space solar-power station makes more sense than one on earth, O'Neill says, because the terrestrial plant gets sunlight only half the day. The space station can convert sunlight to electricity continuously, doubling its efficiency.

O'Neill and the scientists working with him estimate that Island One could be brought into existence for a cost of about $50 or $60 billion—"in the range of one Apollo Project, if inflation is taken into account," O'Neill says—in only ten to twelve years. In its first year of operation, he calculates Island One would return $20 billion in benefits to earth, not only in the form of solar power, but also in other products that are manufactured in space. A good part of the products would be spent building Island Two, Island Three, and so on.

In his book, *High Frontier,* O'Neill pictures an idyllic existence in a space colony. (See Figure 18.) The sphere would turn at two revolutions per minute, providing earthlike gravity for people on the sphere itself. The gravitational force in the sphere would diminish with altitude, so those who climbed the artificial hills would get lighter and lighter. The very center of the sphere would have zero gravity. The problems caused by gravitional variation would be relatively minor, O'Neill says, as would the drift in falling objects due to the spin of the sphere. Coffee poured from a pot to a cup would drift only a millimeter. However, a pop fly that went up thirty-five feet would land six feet away from the spot that a terrestrial infielder would anticipate, so an Island One baseball game would be an interesting experience.

O'Neill sees great advantages in building a new habitat from the ground up. The inhabitants of Island One would not only have their choice of flora and fauna, O'Neill says ("Perhaps, too, we can find less annoying scavengers than the housefly and can take along the useful bees while leaving behind wasps and hornets"), but would also be able to tailor their climate by controlling the amount of sunlight beamed into the colony. The sunlight would be reflected into the sphere by large mirrors aimed at tinted windows that would produce a pleasing facsimile of a blue sky.

Island One residents will probably prefer an eternal summer, O'Neill says. If not, a carefully moderated succession of the seasons can be arranged. Later, when the L5 Lagrange point is occupied by a growing cluster of space colonies, it might be possible to provide a specific season in each colony. The space inhabitant who went swimming in a Hawaii-like colony in the morning might go skiing in an Alpine colony in the afternoon, returning to a suburban-type colony for dinner.

Travel between colonies will be cheap and easy, O'Neill says, because no rockets will be needed. Instead, the spin of the spheres will be used to fling small spacecraft from colony to colony at speeds of hundreds of miles an hour. With an endless supply of energy from the sun, essentially free travel between islands in the sky and all the other benefits of space living, it will not be long before a mass migration from earth begins, O'Neill says. He calculates that more than 7 billion people could be living in space colonies only thirty-five years after the effort starts. And that would be only the beginning. A colony with 10,000 inhabitants could be given a solar electric-propulsion system that would allow them to rove through space almost at will:

> A few decades after the beginnings of the human settlement of space there may well be large groups of people roaming the outer reaches of our solar system, on long-term missions with a scientific purpose. Such groups could be connected intimately to the rest of human society by television and radio, so there would be no reason for them to remain isolated unless they chose isolation for reasons of their own.

O'Neill pictures roving universities, roving laboratories, roving communities of all sorts. However, interstellar travel is not in this picture—at least not until the solar electric system could be replaced by a matter-antimatter engine.

Curiously enough, O'Neill is no enthusiast for SETI. The effect of a signal from a more advanced civilization, he says, will be "to kill our science and our art," from the culture shock of coming in contact with beings who are so much more advanced than the human race. "It seems to me horribly likely that as scientists we would become simply television addicts, contributing nothing of our own pain and work and effort to new discovery." Nevertheless, he does acknowledge that Island One would be "the ideal place" for a search for extraterrestrial intelligence.

The space-colony proposal is closely related to SETI. What O'Neill has sketched is the first step toward a Kardashev Type III civilization. In O'Neill's scheme, moon mining would be only an interim phase in the migration to outer space. Before long, mankind would begin mining the asteroids. From asteroid mining, which would be done by free-roving space colonies, there is a natural progression toward using the material of the major planets, such as

Jupiter, for the purposes of the human race. O'Neill thus is outlining the first steps toward the creation of a Dyson sphere, with the sun at its center and the human race lifted to new heights of technology and affluence by its ability to use almost all of the sun's energy for its own purposes.

O'Neill and the scientists who support him think that the space-colony concept is more than just another interesting idea in technology. They think that it is mankind's only hope. They believe that there is no way to provide decent living conditions to the earth's growing human population by using the resources of this planet alone. Even if the human population somehow stops growing, they say, there is no way in which the earth's resources can be stretched so as to lift everyone to the level now enjoyed by inhabitants of the developed nations. By O'Neill's analysis, the earth does not have the minerals needed for universal prosperity. Energy growth cannot continue indefinitely because the waste heat from energy production will cause dangerous climatic changes. And before long, the earth will run out of living space because there is not enough land to provide decent living conditions if the population continues to grow. Even if the deserts bloom and Antarctica is made habitable, O'Neill says, the earth will be miserably crowded in just a few more generations. In theory, the planets could be made habitable. But planetary engineering is so difficult that the only viable alternative is a move into space colonies.

The O'Neill argument bumps straight into a philosophy that foresees an entirely different future for humanity. Space colonies require complex technology and large amounts of money. The future, as seen by O'Neill, is a high-technology future, in which there are no limits to growth.

But a popular philosophy of the 1970s says that small is beautiful, that complex technology is ultimately destructive, that there are limits to growth and that humanity is pressing against those limits now. The small-is-beautiful philosophers agree with O'Neill that the earth cannot support unending growth. Their solution is not to use technology to transcend the earth's limitations. It is to turn our back on complex technology, to limit our desires, and to place our emphasis on living within the earth's means.

O'Neill sees space colonies that offer humanity the ability to extend its control over more and more of nature. The opposing philosophy sees only one space colony—Spaceship Earth—which will have a stable population, a steady-state economy, a decentralized system based on small-scale technology, and a continual reliance on the same limited supply of recycled resources. In this picture of the future, mankind will not invade the cosmos. It will turn its back on space and will tend its terrestrial garden.

THE CHOICE: SPACE OR STAGNATION

Prediction is a hazardous affair, but it is possible that humanity is at a turning point. O'Neill certainly believes that there is a clear choice between two roads: one leading into space and technological expansion, the other leading toward what he thinks will be stagnation on earth. If such a choice is being made now, the United States is the place where the question will be decided. Only the United States has the grasp of space technology needed to carry out the colonization of space. The Soviet space program is not in the same league. Years after Project Apollo had placed astronauts on the moon several times and Project Skylab had maintained astronauts in orbit for several months, the Soviet Union was still working on rendezvous procedures in earth orbit, with mixed success. Western Europe and Japan are theoretically capable of major space programs, but that capability remains theoretical. In the foreseeable future, either the United States will move toward space colonization, or no one will do it.

The prospect here does not seem bright. In his talks with leaders of the Carter administration, O'Neill found that there was no enthusiasm at all for a major expansion of space technology. Shortly after the Carter administration took control of the National Aeronautics and Space Administration, O'Neill had several meetings with the appropriate officials. He came in asking for billions of dollars. He was offered the hope of a few millions, doled out at a rate that would barely keep the space-colony program alive; instead of hardware, there would be more studies.

The case for a space program in which tens of billions of dollars would have to be spent is by no means clear-cut. O'Neill and his supporters say that it will be possible to move directly from studies and some small-scale models to the prolonged support of dozens, then hundreds, of persons in space for longer periods than have ever been attempted. Their cost estimates are based on smooth fulfillment of extremely ambitious schedules. But the history of past efforts to push technology ahead aggressively shows that the schedules hardly ever are met and that the costs usually are far greater than had been predicted. Project Apollo was an exception because the moon program was not really an attempt to break new technological ground, at least to any great extent. O'Neill and his supporters may be correct in all their estimates, but there are grounds for doubt.

There would have been less doubt in the 1950s or 1960s, when money was much freer and doubts about technology were fewer. The O'Neill request for billions of dollars must be seen in the context of the 1970s. The Space Shuttle, on which all of the space-colony plans are based, has just managed to survive several close calls in Congress, and its future is placed in doubt with every new federal budget. As for support for space colonization from the astronomy community, astronomers generally are much more concerned about a project called the Space Telescope, whose costs are measured in millions, not billions. It is a symptom of the times that the name of the project was changed quietly from Large Space Telescope; the word "large" was dropped to soothe congressional feelings. Despite some reductions in the scope of the Space Telescope project and despite assiduous lobbying, the orbiting telescope has been postponed several times. Complete cancellation is a constant danger. A scientific community that must work hard to maintain federal funding for what are regarded as essential research efforts is in no position to talk in terms of many billions of dollars for a somewhat risky effort to explore a new space frontier.

The Carter administration is not enamored of new frontiers; its decision makers appear to have embraced the steady-state philosophy. All the administration's decisions on energy are based on the principle that the future will see a drastic reduction in the energy growth rate. And while the space program is not fading away, there is a distinct lack of the enthusiasm that John Kennedy and Lyndon Johnson showed for space exploration. To try to get things moving, O'Neill has set up a nonprofit Space Studies Institute, which solicits public contributions, and is lobbying Congress for support. But it is hard to see how public contributions and congressional resolutions can be translated into billions of dollars of federal appropriations when the administration is at best indifferent to a program and could be actively hostile.

However, things could change. Administrations are not in power forever. In practical terms, not much can be done about space colonies until the Shuttle has flown a decent number of orbital missions, which will not happen until the early 1980s. Public opinion may change markedly over the years; it certainly has changed since the enthusiastic days of the 1960s when Project Apollo was made a national cause. Perhaps the decision to be made about the space colony scheme is a milestone; perhaps it will mark the moment when humanity either goes down the road toward continued technological expansion or takes the route to a self-contained, steady-state existence. But the decision has not yet been made. The choice is still open. It may remain open for many more years.

There is a more pressing decision of immediate importance to SETI—a decision about radio frequencies. The water hole is a popular band for a number of different transmissions. The fact that false alarms have occurred continually even in the limited number of searches that have been made for extraterrestrial intelligence indicates just how popular it is. Unless something is done to limit the use of frequencies in the region of the water hole for everyday civilian and military purposes, a search for intelligent signals may become impossible for earth-based antennas. It may be necessary to put SETI antennas in space to avoid interference, which would mean an enormous increase in cost.

The SETI community is very much aware of the problem. In 1979 a World Administrative Radio Conference will be held to allocate uses for different bands of the radio spectrum. The decisions made at the conference will probably govern the use of the spectrum for the rest of the century. When the convention meets, it will have before it a proposal to study restrictions on the most likely SETI frequencies. At present, the bands from 1,400 to 1,427 MHz are allocated to radio astronomy use only. The bands from 1,427 to 1,727 MHz, which lie between the hydrogen and the hydroxyl lines, are also important to SETI. An effort will be made to protect the whole region from 1,400 to 1,727 MHz for SETI. A resolution has been prepared which does not name the specific frequencies to be protected. But it does say that "it is believed to be technically possible to receive radio signals from extraterrestrial civilizations," and it asks the conference to determine the preferred frequency bands to be searched and "the preferred locations, on earth and in space, for a search system."

Unlike the space-colony proposal, which could reasonably be described as visionary, the idea of protecting radio frequencies for SETI is determinedly practical. There are only a limited number of usable radio frequencies, and the demands on those frequencies keep growing. The response of the world conference to the request from the SETI community will thus be a valuable indicator of the status of the search for extraterrestrial life today. Is SETI important enough to push aside some powerful financial and national interests? Are commercial interests and military establishments willing to undergo some inconvenience for the sake of an admittedly small chance of making contact with another intelligent civilization?

THE POETRY OF SETI

These questions do not reflect the poetry that has been the driving force behind SETI until now. Yes, poetry is the right word. As we read the early writings on extraterrestrial communication—and early means anything that dates back about ten years—there is a clear feeling of wonder that can be described as poetic even though it is clothed in scientific language.

There is speculation that Phobos and Deimos, the two moons of Mars, might be giant artificial satellites (they are, as photographs from the Viking orbiters have shown, pockmarked lumps of rock). There is speculation that the color changes on Mars visible through terrestrial telescopes might be due to vegetation (they are caused by dust storms). There are Kardashev civilizations and Dyson spheres, asteroids to be mined and comets to be colonized. Now we are down to haggling about money for hardware and the reservation of radio frequencies for a search that could last for decades.

In other words, the search for extraterrestrial intelligence is coming of age. It is moving out of its poetic adolescence at what seems to be an unfortunate time, a time when hard-skeptical questions are the rule. However, those questions would be asked sooner or later even if circumstances were different. If a search for extraterrestrial signals had started in the 1960s, the chances are small that it would have succeeded by now. We might have government officials asking impatiently whether spending would have to go on at the same level, and how long the government might have to spend money on a fruitless effort. We probably would have the skeptics citing the familiar numbers to support the view that, in the words of D. R. Bates:

> It would be improvident for us to proceed on the assumption that a government of a civilization belonging to the hypothetical communication network would finance a prolonged beacon effort aimed merely at hastening contact with any relatively recently developed civilization rather than leaving the initiative to the newcomer as the one having the greater incentive and the easier problem.

There is an answer. "We believe that signals have been falling on the earth for many millions or billions of years," said John Billingham. "If that is the case, why not look for them here. Perhaps if we are sufficiently vigorous about it, we stand a good chance of success."

So far, the view of Billingham and other believers in the search for extraterrestrial intelligence has prevailed. The two major technological powers on earth, the Soviet Union and the United States, are both committed to projects to listen for signals from another civilization. The size of the projects and the specific strategies that have been adopted may be disappointing to some people in the SETI community, but the important thing is that the commitment has been made.

THE FIRST CHAPTER IN A LONG STORY

However, this is only the beginning. It is the first chapter in what must inevitably be a long story. Realistically, there is little chance that the limited listening programs now underway or being planned will succeed in making contact. The odds are great that we will be asked not only to continue these programs but also to expand them despite years of negative results. Ten and twenty years from now, the same issues will confront the decision makers—except that they will have to make their decisions after many years in which no signals from other civilizations have been detected. It is reasonable to predict that the really difficult decisions about the search for life elsewhere in the galaxy will not be made until the twenty-first century. By then, if present trends hold, we will have had more than two decades of a continuing search for extraterrestrial signals, on a modest but growing scale. Everything that is easy will have been done. Several all-sky scans at relatively low sensitivity will have been completed. Probably someone will have looked at all the nearby sunlike stars in all the obvious frequencies of the water hole. At that point, a decision will have to be made. The choice will be between giving up, doing more of the same, or paying for a major expansion of listening programs.

If this is in fact the scenario that is played out, the decision that will be made in the year 2000 will be one of the most significant in human history. It will be significant because the search for extraterrestrial signals is more than just another scientific and technological exercise. It is one of the measures by which we can judge the prospects for the future of mankind.

Scientists cannot measure something without changing what they measure. The chemist who measures the temperature of a solution changes that temperature by inserting the thermometer. In just the same way, the decision about continuing a search for extraterrestrial intelligence affects the odds that the search will succeed. By the principle of mediocrity, we must say that earth is not basically different from other planets where technology may have arisen. Like the earth, those other planets are assumed to have all the problems that afflict our society: pollution, overpopulation, and the rest of the troubles of our time. They must also have come to crossroads where the choice had to be made between embracing technology fully or turning away from technology. They must also have had to decide about continuing an effort to make contact with other intelligent species or abandoning that effort.

The only way we have of even hazarding a guess about the decisions that have been made elsewhere in the galaxy—barring the lucky stroke of making contact quickly, in the next few years—is to observe how those decisions are made on earth. Thus, the measurement influences what is being measured. To give up the search for other civilizations is the most direct way to ensure that we will remain alone. To continue the search is the only way to keep alive hope for making contact with another intelligent species.

In the minds of the people in the SETI community, that is perhaps the most important hope available to the human race. Success in the search for another advanced civilization offers much more than the hope for a fascinating conversation. It offers the hope that civilizations such as ours can break out of their times of trouble into a long era of stability and prosperity. By the formula invented by Frank Drake, the search for another civilization can succeed only if there is a large value of L, the lifetime of an advanced civilization. A decision to continue the search for such civilizations is thus a vote for a large value of L. At a time when despair is common, it would be a vote of confidence in the future of the human race and in the ability of humanity to master technology.

Whatever the search for intelligence in the universe may find in the future, its most important role today is in keeping that hope alive.

INDEX